D0990894

INEQUALITY IN THE UK

INEQUALITY IN THE UK

ALISSA GOODMAN
PAUL JOHNSON
STEVEN WEBB

OXFORD UNIVERSITY PRESS

1997

339.2
G65i

Oxford University Press, Great Clarendon Street, Oxford OX2 6DP
Oxford New York
Athens Auckland Bangkok Bogota Bombay Buenos Aires
Calcutta Cape Town Dar es Salaam Delhi
Florence Hong Kong Istanbul Karachi
Kuala Lumpur Madras Madrid Melbourne
Mexico City Nairobi Paris Singapore
Taipei Tokyo Toronto
and associated companies in
Berlin Ibadan

Oxford is a trade mark of Oxford University Press

Published in the United States
by Oxford University Press Inc. New York

© Alissa Goodman, Paul Johnson, and Steven Webb 1997

All rights reserved. No part of this publication may be reproduced,
stored in a retrieval system, or transmitted, in any form or by any means,
without the prior permission in writing of Oxford University Press.
Within the UK, exceptions are allowed in respect of any fair dealing for the
purpose of research or private study, or criticism or review, as permitted
under the Copyright, Designs and Patents Act, 1988, or in the case of
reprographic reproduction in accordance with the terms of the licences
issued by the Copyright Licensing Agency. Enquiries concerning
reproduction outside these terms and in other countries should be
sent to the Rights Department, Oxford University Press,
at the address above

This book is sold subject to the condition that it shall not, by way
of trade or otherwise, be lent, re-sold, hired out or otherwise circulated
without the publisher's prior consent in any form of binding or cover
other than that in which it is published and without a similar condition
including this condition being imposed on the subsequent purchaser

British Library Cataloguing in Publication Data
Data available

Library of Congress Cataloging in Publication Data
Goodman, Alissa.
Income inequality in the UK/Alissa Goodman, Paul Johnson,
Steven Webb.
Includes bibliographical references (p.   ) and index.
1. Income distribution—Great Britain. 2. Income distribution—
Great Britain—Statistics. I. Johnson, Paul, II. Webb, Steven.
III. Title.
HC260.I5G58   1996   96–32062
339.2'0941—dc20

ISBN 0–19–877360–9
ISBN 0–19–877361–7 (pbk.)

Typeset by J&L Composition Ltd, Filey, North Yorkshire
Printed in Great Britain
on acid-free paper by
Bookcraft (Bath) Ltd., Midsomer Norton, Somerset

PREFACE

The work on which this book is based has taken place at the Institute for Fiscal Studies (IFS) over a period of at least eight years. Over that period, we have worked with a number of colleagues on the income distribution, and their work has been important in developing the data and techniques that form the central part of this volume. For this, thanks are due to many colleagues, but special recognition, for their work in areas that have contributed to this book, is due to Chris Giles, Amanda Gosling, and Graham Stark.

Outside IFS, many people have helped us in our work and have been instrumental in taking forward the empirical analysis of the income distribution in the UK. Special recognition is due to analysts within the Department of Social Security (principally in ASD6, headed by Gordon Harris) who have been at the forefront in designing and computing accurate, reliable, and digestible figures on the UK income distribution. In the academic community, a number of individuals have stimulated our interest with their work on the subject of inequality—John Hills and Stephen Jenkins deserve a particular mention in this respect.

The funding needed for the writing of this book was provided by the Economic and Social Research Council (ESRC) as part of its support for the ESRC Centre for the Microeconomic Analysis of Fiscal Policy at IFS. Thanks are also due to the Joseph Rowntree Foundation, the Department of Social Security, and the House of Commons Social Security Select Committee for funding many of the original analyses underlying this publication.

An empirical analysis such as this requires the use of a great deal of data. We are grateful to the Central Statistical Office for providing the Family Expenditure Survey data on which many of the analyses are based. The ESRC Data Archive at Essex provided British Household Panel Survey data. We bear full responsibility for all analysis and interpretation of these data.

Andrew Dilnot and Richard Disney have put considerable effort into reading and commenting upon various drafts of this book, and for that we thank them.

University Libraries
Carnegie Mellon University
Pittsburgh PA 15213-3890

Finally, Judith Payne is owed a special debt of gratitude for her usual swift and astonishingly careful and accurate copy-editing and proof-reading.

CONTENTS

TABLES

FIGURES

Introduction

How well off am I relative to my peers? How poor are the poor? How much better off are the rich now compared with twenty years ago? Is the gap between rich and poor growing?

Such questions are common currency in the political discourse. They are central to the recurring debates over tax policy and the effects of budgetary announcements. On the answers to such questions depends any intelligent analysis of the scope and structure of the welfare state. It is the purpose of this book to set out comprehensive answers to these questions and many more by analysing the distribution of income, and also of spending. It describes the distributions, how they are generated, and how and why they have changed.

We introduce the discussion by considering some of the reasons for thinking that the distributions of income and spending might matter and some of the reasons for the writing of this volume. We then give the briefest of indications of the contents of the ten chapters that constitute the central part of this work.

WHY WE NEED TO KNOW

The way in which income and spending are distributed among individuals matters for all sorts of reasons. It matters for politics, for understanding political processes and outcomes. It matters in informing the policy decisions of politicians and other policy-makers, obviously in the fields of tax and social security, but also for health, education, and urban policy. It matters to economists in understanding how the economy works at both the micro and macro levels.

If J. K. Galbraith is to be believed, then the existence of an adequately large group who are comfortably off might be a sufficient condition for the continued existence and success of a political agenda that neglects the interests of a (substantial) minority who are far from comfortable. Some commentators[1] see the degree of inequality in the UK as both a result of, and a buttress to, particular problems of the British economic and political system. Others interpret similar facts in a positive light as providing incentives to 'get on' or as reasons for radical changes in tax and social security to improve those incentives. One does not have to take sides in this sort of debate to see the importance of understanding how incomes are distributed.

While there is certainly debate in the political arena about whether it is a good or a bad thing that there is a large gap between the rich and poor, there is no question that an accurate knowledge of the extent of that gap and reasons for it are central to understanding, or at least interpreting, political processes.

Fairness

But does the level of inequality matter *per se*, from the point of view of fairness? If, for example, one group of the population received the same amount now in real (price-adjusted) terms as they did twenty years ago, would it be a matter of any concern if they received less relative to everybody else?

In its simplest and least controversial sense, the term 'income inequality' is used to describe the fact that some people receive higher incomes than others. The term in this sense could equally well be applied to differences in these people's weight, or height, or shoe size. In its common usage, however, the term 'inequality' is often used not just to refer to pure differences, but to embody some sort of value judgement about how *fairly* things have been distributed. A good example of this is the *Daily Mirror*'s assessment of some recent research: 'So the rich are getting richer and the poor are missing out. That comes as no surprise, though the figures published yesterday show an accelerating inequality which is deeply offensive. Not just politically offensive, but offensive to that very old-fashioned, very British notion, *fairness*' (our italics).[2]

People coming from different ethical viewpoints will judge statements such as that differently. One could claim that there is no interesting ethical content to the level of income or wealth enjoyed

---

[1] Will Hutton (1995), in his popular book *The State We're In*, puts this thesis forward particularly powerfully.

[2] *Daily Mirror*, 3 June 1994, commenting on Goodman and Webb (1994) and Jenkins (1994b).

by anyone at any time, and therefore in its distribution among individuals. In particular, it is possible to argue that all that matters from an ethical point of view is how people came to have their money, not the level of their income or how it relates to the incomes of others. If a distribution is arrived at through the free workings of the market, through contracts freely entered into, then, the argument goes, the distributional outcome is ethically uninteresting.[3]

One would not have to agree that the process is *all* that matters to see that what leads people to have the income that they have is important. We might well take a different view of a situation in which the inequality in incomes comes directly from the different degrees of effort made by the individuals concerned, from the view we might take of a situation in which incomes are effectively determined through chance of birth. This is why it is so important to understand what causes inequality and changes in inequality, and why much of what follows concentrates on such issues.

Policy Formation

A proper understanding of the distribution of income is crucial for many areas of government policy. Titmuss (1962, p. 17) was passionate about the need for the government to shape policy on the basis of proper information about the distribution of income: 'Chancellors of the Exchequer . . . need more facts in the framing of social and fiscal policies as well as for the general oversight of the economy. . . . The days are long since past when policies for health, education, and social security could be shaped without reference to trends in the distribution of income and wealth'.

The amount that any tax on income will raise depends on the numbers of people affected and the incomes they have. One cannot begin to assess the structure or effects or even the progressivity of any income tax system without a knowledge of the income structure of the population on which the tax is being levied. High rates of tax on a very small number of people are likely to be ineffective as revenue-raising measures. High rates on large numbers of people might well be too politically sensitive to implement.

The appropriate design for any social security system is also intimately bound up with the income distribution. The appropriate level

---

[3] The classic rendering of this sort of position is that of Nozick (1974).

at which to set benefits is likely to be affected by the incomes of non-benefit-recipients. One would not generally want out-of-work benefits to move recipients above workers in the income distribution. Affecting large numbers of people by high marginal withdrawal rates is generally something to be avoided where possible.

There is a long and often fruitless debate about the structure of social security benefits. It has often focused on the relative merits of means-tested (income-related) benefits such as Income Support, which go only to those with incomes below a certain level, and 'universal' benefits such as Child Benefit or the Retirement Pension, which go to all those with a particular set of characteristics (having a child or being over state pension age in these two cases). Probably the central issue in this debate ought to be an assessment of the income levels of those likely to be affected by the benefits. If a group (such as pensioners) can be identified who have uniformly low incomes, then means-testing is likely to be unnecessarily complex and expensive. This is especially true where there is a serious potential for causing disincentives either to work or to save. On the other hand, where a group has diverse incomes, universal benefits might well be relatively wasteful and unnecessary. One certainly needs detailed information on the incomes of these groups in order to arrive at an informed policy prescription.

Many other aspects of government policy have important implications for the distribution of income. Government policy towards unions and pay-setting will affect the distribution of earnings, which is an important component in the distribution of income. Government policy towards education and training may affect individuals' ability to command earnings and therefore affect the patterns of incomes in the economy.

In the field of overarching macroeconomic policy, some authors have proposed a relationship between the degree of inequality in a country and its macroeconomic performance.[4] On the one side, it is argued that the inequality is related to flexibility and to work incentives—people work harder because the penalty for not doing so is penury and rewards for doing so are great—while on the other side, it is argued that inequality promotes instability.

----

[4] See, for example, Corry and Glyn (1994).

WHY THIS BOOK NOW?

That a knowledge of the type and degree of inequality that exists is evidently important is not of itself reason enough to devote a new book to its exposition.

As far as we know, the last self-contained book devoted to a comprehensive analysis of inequality in the UK was that of Atkinson (1983).[5] Much has happened to the distribution of income since then. There has been a burgeoning in the literature on the UK income distribution, and giant strides have been made in the use and availability of data. This book presents, analyses, and explains this new information.

There has also been an increasing concern from government, academics, and bodies interested in public policy in the whole area of inequality. An introduction to a book of this sort would not be complete without a recognition of the work done by analysts at the Department of Social Security in setting the highest of standards in the preparation and presentation of inequality statistics which are now published annually. Their publication indicates a genuine interest within government in inequality. In addition, the Joseph Rowntree Foundation threw its weight behind the inequality debate, and a committee set up by it published a document outlining its concerns at the beginning of 1995. The report received unprecedented public interest and coverage. Just a year before that, the Commission on Social Justice, set up by the late John Smith, took as one of its central concerns the degree of inequality persisting in the UK.

So to motivate this new volume is not hard. Let us briefly review some of the major changes that occurred during the 1980s and early 1990s.

The first thing that is new is the unprecedented increase in income inequality over the 1980s. When Atkinson wrote *The Economics of Inequality*, he had to concentrate on trying to discern if there had been any unambiguously identifiable changes in the income distribution over the preceding years, and on why the distribution appeared to be relatively stable. He was bleak about the prospect for making unequi-

---

[5] Although we use Atkinson's seminal work as the point of departure here, this volume has significantly different intentions from his. We do not consider the US as well as the UK; we do not look at the distribution of wealth; we spend less time on earnings and the determinants of earnings than he did; and we do not consider reforms of tax and social security policy as a means of changing the distribution.

vocal judgements about trends in the income distribution. In conclud-
ing his chapter on the empirical evidence about the distribution of
income in the UK, he comments 'Ideally, we would end this chapter
by drawing together the different elements and reaching a clear-cut
conclusion about the path of income inequality in Britain . . . At the
same time, it will be apparent that this goal is some way off' (Atkinson, 1983, p. 94). We have the much more fertile ground of describing
a massive and unambiguous widening in the income distribution and
attempting to explain it.

We also have access to a long time-series of high-quality datasets
which allow comparisons to be made over a period of more than thirty
years. The data are detailed enough to allow the constituent parts of
income to be disentangled and the family circumstances of people at
different parts of the distribution to be described. The availability of
the data is combined with new levels of computing power which
allow this detailed analysis to be carried out much more easily than
in the past.

Work looking at how the distribution has changed and why has now
reached a point at which a substantial literature and even greater array
of data need to be brought together and assessed. But more recent
analysis is already casting a new light on the whole issue. Whereas a
book of this type written only two or three years ago would probably
have used just cross-sectional data and would have stopped at the
claim that income inequality has unambiguously increased, we have
been able to extend the analysis in two important ways which both
enlighten the discussion but also serve to temper some of the conclusions.

The first of those new sets of analysis relates to the distribution of
spending. As is made clear in Chapter 1, on the methodological issues
surrounding the measurement of inequality, descriptions of the distribution of spending or consumption add much to our understanding
of the distribution of economic welfare. Indeed, many economists
have argued that spending provides a better measure of living standards than does income. The pattern of change in spending inequality
has been considerably less dramatic than the change in income
inequality. While it does not change the general conclusion that there
has been a major widening in the distribution, it certainly casts some
doubt on the real extent of that change. It also suggests that a part of
the increase in income inequality is a result of an increase in the
uncertainty associated with incomes. It is perhaps nowadays more

likely that individuals will see their incomes rising and falling more frequently and more violently than was the case in the past.

This possibility is dealt with directly in Chapter 9, which introduces a second novel part of the distributional picture. Only since 1994 have we in the UK had access to panel data covering a sample of the full population allowing us to look at how individuals' incomes change from year to year. In the past, it has, for example, been possible to compare the incomes of the poorest 10 per cent in one year with those of the poorest 10 per cent in another year, but it has not been possible to say what happens over time to those who find themselves in any particular part of the income distribution. What we show in Chapter 9, using just the very first couple of years of available panel data, is that there is a great deal of movement from one part of the income distribution to another even over a relatively short period of time. This fact should be borne in mind when it comes to interpreting the main findings on how the income distribution has changed.

THE REST OF THE BOOK

In describing the motivation for this work, we have already touched upon some of the main ingredients that go to make up this volume. But in this brief section, we set out a more coherent guide for the reader.

Chapter 1 provides a conceptual and methodological framework for the analysis of inequality. Its main concern is less with a discussion of what inequality is than with the question 'inequality of what?'. Even after settling on either income or expenditure, one has to decide exactly what counts as income—for example, whose income it is that should be compared and how to deal with the fact that people live in households and not just independently and alone.

Chapter 2 describes the current UK distribution of income—how many people there are at different income levels, what sorts of people these are, where they get their income from, and how unequal the distribution is overall. Chapter 3 follows on from this, looking at how these things have changed over time. The shape of the distribution is very different today from what it used to be. Incomes are now much more unequal than they were even ten years ago, and the nature and causes of low incomes have changed. This chapter looks at these issues in detail.

Instead of looking at incomes, Chapter 4 considers inequality of spending, how it differs from income inequality, and what the differences tell us about trends in living standards.

The following three chapters analyse some of the forces that lie behind the changes in inequality. The first, Chapter 5, breaks income down into its constituent parts, considers from where people at various points in the income distribution receive their money, and how the importance of the various constituent parts has altered over time. Particular importance is attached to the largest of the income sources—that is, earnings. Chapter 6 then moves on to consider the role of demographic and labour market forces in shaping the trends in inequality. The period of our study has seen major changes in the demographic structure of the population, in patterns of family formation and dissolution, and also in the labour market.

Chapter 7 goes on to examine the trends in income inequality further, by looking at the changes in the tax and benefit system over the past three decades and the effects that these have had on income inequality. It looks at the main sorts of direct taxes and benefits and at who pays and receives them.

Chapter 8 focuses on an important aspect of low incomes: poverty. It looks at different definitions of poverty, and how these affect the number of people who are deemed to be poor. In the following chapter, we look at a hitherto much neglected aspect of the income distribution, and one which is particularly relevant for the analysis of poverty: income dynamics. How long do different sorts of people spend on low incomes, and what causes people to move into and out of low incomes? These are issues addressed in Chapter 9.

The book ends by drawing together some of the threads from the analysis and setting out the most important conclusions.

Finally, a word about two issues that are not covered in the following chapters. First of all, this is a book about the UK. We make almost no reference to what has happened elsewhere.[6] That we have done this reflects in part a serious problem in putting together fully comparable sets of data and in interpreting results from cross-national studies that have been attempted. Work in this area is nevertheless improving all the time and should be one aspect of distributional analysis that will take major leaps forward.

---

[6] The interested reader is referred to Gardiner (1994) and Atkinson, Rainwater, and Smeeding (1995).

A second issue that we do not directly address is that of wealth inequality. The state of statistics on wealth holdings in the UK is far from what one might desire. What we do know is that wealth is distributed massively more unequally than is income. That distribution is also likely to have changed over the 1980s with the spread of owner-occupation, rising house prices, and increased coverage and value of private pensions. For a discussion of the general picture on wealth holdings, the reader is again referred to Atkinson's 1983 book and also to Atkinson, Gordon, and Harrison (1989) and Feinstein (1996).

1   Measurement Issues

INTRODUCTION

Before proceeding with the empirical analysis which forms the core of this work, it is important to be clear about a number of issues both of principle and of methodology. Even before considering how best to describe different distributions and what it is we mean by inequality, a whole host of questions arise which can perhaps best be summarised by the question 'the distribution of what?'.

Ideally, one would want to compare the living standard of each individual in the country. Sadly, we cannot measure this. The concept of 'standard of living' is a slippery one, and there is certainly no agreement among economists or other social scientists regarding what constitutes a standard of living. The first section of this chapter deals with this question and in particular looks at the relative merits of income and expenditure as proxies for living standard.

But then a multitude of questions arise regarding what is counted as income or spending, the period over which they should be measured, and the treatment of a whole series of other pieces of information. These are addressed in the second section of the chapter.

The third set of issues relate to whose income or spending is of interest. If we are interested in individuals, how do we take account of the fact that individuals live in families and to some extent share their incomes? How do we know to what extent they do share? How should we compare the incomes or spending of families of different sizes?

Practical answers to all these questions will depend on the information available. For this reason, the fourth part of the chapter discusses some of the issues raised by the fact that imperfect data have to be used in constructing pictures of income and spending and inequality. We describe the main attributes of some of the main sources of data that we do use.

Finally, some consideration is given to the measurement of inequality itself. There is a proliferation of measures with various attributes. In the following chapters, we make use of just a small number of them.[1] The main features of these measures are described in the final part of this chapter.

LIVING STANDARDS

Let us imagine that we are trying to measure the living standard of a single individual with no family to take into account and no other problems regarding sharing. What is it that we should try to measure?

Welfare or living standard as such is not readily measurable. But that is no reason for despair. There are plenty of things that we can measure which tell us much about living standards—income, spending, wealth, and health, for example. Each of these is likely to reveal something about overall living standard, though without providing a full picture. As Sen has reflected,

Within the general notion of the living standard, divergent and rival views of the goodness of life co-exist in an unsorted bundle. There are many fundamentally different ways of seeing the quality of living, and quite a few of them have some immediate plausibility. You could be *well off*, without being *well*. You could be *well*, without being able to lead the life you *wanted*. You could have got the life you *wanted*, without being *happy*. You could be *happy*, without having much *freedom*. You could have a good deal of freedom without achieving much. (Sen, 1985, p. 1.)

The make-up of the notion of well-being is rich and varied. Nothing we can do will provide a full description of it. There are two magnitudes that are used in this book—income and expenditure. In order to understand why, we start by considering the relationship between income, consumption, and wealth.

Income, Consumption, and Wealth

Income and consumption differ from concepts of welfare in that they tell us about command over resources—potential in the case of income and actual in the case of consumption. They do not tell us about welfare *per se*. They are only perfect or ideal proxies for living standards if we assume that living standards depend only on material

---

[1] For a comprehensive discussion of the measurement of inequality, the interested reader is referred to, for example, Morris and Preston (1986), Lambert (1989), and Cowell (1995).

consumption. A living standard so defined is a much more limited concept than that outlined by Sen above. It takes no account of health or freedom or achievement. It does not measure happiness unless happiness is directly equivalent to level of income or consumption, which of course it is not. It does not measure the 'worth' of an individual. Income and consumption are economic measures of resources.

There is one other magnitude that is likely to be of economic interest (and capable of measurement) and that is wealth. So we have income, consumption, and wealth. What is each of these, what is the relationship between them, and what does each tell us?

Life-Cycle Theories

Let us return to our isolated individual. Suppose we observe him over a period of a year. The relationship between his income, consumption, and wealth should be clear. He may receive income from a number of sources—earnings, investments, social security. The value of any wealth he holds might increase, and this increase in value should be considered as part of his income. Out of this income he can save. The difference between his income and his saving is his consumption. If over the period of the year his income is greater than his consumption, then his wealth at the end of the year will be greater than his wealth at the start of the year. Of course, his consumption might be greater than his income. In that case, he could be running down his wealth, or he could be borrowing. If he ends the year in debt, he would have negative net wealth. These relationships can be summarised by the following very simple equations:

$$W_1 = W_0 + I - C$$

or

$$W_1 = W_0 + S$$

where W_0 represents his wealth at the start of the year, W_1 is his wealth at the end of the year, I is his income over the year, C is his consumption, and S is savings.

If we now consider the same person and his consumption, income, and wealth over a longer period—a lifetime perhaps—we might get a clearer idea of what each is telling us.

According to economic theories such as the permanent income hypothesis (Friedman, 1957) and theories of the life cycle (for example, Ando and Modigliani (1963)), people smooth their consumption when their incomes vary. This is because they base their consumption decisions not just on the current income that they are receiving at any particular point in time, but on their best guess of what the longer-run resources available to them are. For instance, students on very low incomes borrow in the expectation of higher incomes in the future, and many working people save in the form of a pension to provide income for themselves in their old age.

Suppose our isolated individual enters his adult life with nothing. He then goes through three periods—education, work, and retirement. He only earns any money in the second, working, period. Let us assume that there are no student grants or state-provided old-age pensions. Life-cycle theories would predict that he would borrow during the period of education. In the working period, he would pay off these debts and save for retirement. In retirement, he would receive some investment income from his assets but he would also use up his assets. He would die (leaving nothing as he has no children to inherit from him) with nothing.

How would his consumption, income, and wealth change over time? The first major prediction of life-cycle theories is that his consumption would not actually change very much over time. Assuming access to financial institutions through which he can borrow and lend, he should roughly speaking smooth his consumption over his lifetime. (The other assumption required for this to hold true is that there should be diminishing marginal utility of consumption. In other words, I gain less utility by spending another pound than I would lose from spending one pound less. So if my spending were not equal across periods, I could always increase my overall utility by raising spending in the low-spending periods and reducing it in the high-spending periods, until they were equal.)

At the start of period 1 (his student years), our isolated individual has no wealth. During that period, he has no income. At the end of the period, his wealth is negative. His consumption during the period, clearly, is greater than his (non-existent) income. Come period 2, he starts work. During the period, he pays off his debt and starts saving for retirement. So his wealth gradually becomes positive and grows throughout this period. His income is higher than his consumption because some of it is going on paying back debts and on saving. By the beginning of period 3, he has substantial wealth, but this is run down to zero by the end of the period (his death). The wealth creates some income but this is clearly lower than his consumption for which he can dissave.

How plausible is this sort of description and what does it mean for the measurement choice and for the measurement of inequality?

Plainly, people do not act exactly like this. Capital markets are not perfect and it is not always possible to borrow as much as one might want. There is a good deal of uncertainty in life, so saving might be higher than necessary because of worries over unforeseen eventualities, or lower because of myopic expectations. People do not know when they will die. They might very well want relatives to inherit money.

There are a thousand and one objections to the model in its purest form. But the general predictions are perfectly reasonable. Students borrow on the expectation of future earnings. Workers save in pensions and in other ways to provide for themselves in retirement. Pensioners live off these savings.

Suppose, as we should, that we accept that there is some truth in this model; what would it tell us about the uses of income, consumption, and wealth as bases for measurement? The first message is that consumption should provide a better indication of permanent or lifetime well-being. The fact that students have very low incomes does not tell us anything about their lifetime status. Where they are able to borrow, it may not tell us much about their current well-being. Similar arguments are applicable to wealth. In this simple model, the path of wealth followed by the individual is from nil to negative to positive and back to nil again. But this does not indicate any lower utility at times of low wealth than at times of high wealth; wealth here is used purely as a means of smoothing consumption over time.

The corollary of this, of course, is that with a population greater than one, and at different points in their life cycles, income, consumption, and wealth will give rather different measures of inequality. Suppose there are three individuals identical in all respects except age. The first has just started at college, the second has just started work, and the third has just retired. Even though they might be identical in every respect but their age, an income-based inequality measure will show significant inequality. Person 1 will have no income, person 2 will have a significant amount, and person 3 just a little. A wealth-based measure would rank the just-retired individual (person 3) as best off followed by person 1, and person 2, best off on the income measure, would be worst off on the wealth measure. A consumption measure would find no inequality.

If the real world conformed to the assumptions behind this model, there would be little question that consumption would be the appropriate measure for use in exploring inequality. But while it is vital to bear these ideas in mind, the real world is sufficiently different from this for income distributions (and indeed wealth distributions) to be of interest as well.

The fact that capital markets are not perfect is important. Inherited wealth matters. Individuals are both myopic and generally risk averse. Expenditure, which we can measure, is not the same as consumption, which is what we would ideally like to measure.

Command over Resources

Income and consumption emphasise different aspects of households' financial situations. One reason why income has often been chosen over consumption as a measure of the standard of living is that it is said to measure the household's *potential* living standard, whether or not the individual actually chooses to achieve it. As it is expressed in DSS (1995, p. 10), the aim of the Department of Social Security's official (income-based) low income statistics is 'to measure people's potential living standards derived from goods and services financed from disposable income'. Another common way of putting this is that income (along with wealth) measures the *command over resources*. This accords well with a basic intuition about living standards, which is illustrated by the following example.

Imagine there were two identical individuals (of the same age, for example) with exactly the same income. If one spends all his income, whereas the other decides to save some, the income measure gives them both the same standard of living. The income measure tells us that they both *could have* achieved the same level of expenditure on goods and services, if they had chosen to. Someone receiving a million pounds a year might decide to spend next to nothing out of his vast income. This is an obvious situation where it is potential rather than actual enjoyment of goods and services that we would want to compare against others.

The argument here is that income is a better measure because it gets at potential consumption, rather than consumption itself, which is a matter of choice. But income only truly reflects potential consumption if households cannot live off their former savings or borrow. If they can, then their potential consumption at a particular point in time depends not just on their current income, but on the amount they have saved in the past and the amount they can raise by borrowing.

Consider a different two households. One is the million-pounds-a-year man who has now reached pension age and whose income has all but dried up. Fortunately for him, after years of penny-pinching, he has amassed an enormous stockpile of money under his mattress. He now has the free time to go out and enjoy the good life for the first time, buying all the luxuries he has denied himself for so long. The person we are comparing him with has also just retired, but without the benefit of any savings, and, unable to borrow, he has to live on just his state pension and Income Support. The incomes of these two might be similar, but their consumption levels diverge dramatically. In this situation, there seems to be a strong case for saying that it is the differences in consumption levels and not in incomes which we should use to compare living standards.

Whereas it is the income measure which can take on board the fact that if households save, it is their own choice, it is the consumption measure which reflects the choice to borrow or to run down savings. Although consumption measures might better reflect the longer-run resources available to the household,[2] current income can better reflect the immediate predicament in which a household finds itself. But in the presence of accumulated savings or wealth, income might not even reflect this immediate predicament because it does not give a full view of command over resources.

Effects of the Choice

The choice between income and expenditure as the measure of household living standards does in fact make quite a difference to the picture of inequality that is seen or presented.[3] For example, empirical research demonstrates that the position of pensioners relative to the rest of the population is worse if living standards are measured by

---

[2] Although the argument that comparisons of current (as opposed to lifetime) income across people of different ages reflect differences in their immediate circumstances but not necessarily their lifetime welfare also applies to current expenditure; see Blundell and Preston (1995).

[3] Some studies have directly compared the distribution of income and the distribution of expenditure, and changes over time. Blundell and Preston (1995), DSS (1995), Goodman and Webb (1995), and Smeaton and Hancock (1995) discuss differences in income and expenditure in the UK; Cutler and Katz (1992) and Slesnick (1992) look at the US experience.

expenditure rather than by income. Pensioners' expenditure tends to be low relative to their income by comparison with the rest of the population (which does itself appear to be at variance with the stylised predictions of the simple life-cycle hypothesis).

The self-employed are another group in the population whose incomes and expenditures provide very different stories about their standards of living. Although there are many self-employed on very low incomes, there are far fewer who have very low spending. As a group, the self-employed are over-represented at the bottom of the income distribution. They make up around 10 per cent of the population but over 15 per cent of the poorest tenth of the income distribution. On the other hand, they are under-represented at the bottom of the expenditure distribution, making up only around 5 per cent of the poorest expenditure tenth. One possible explanation for this is that the incomes of the self-employed are more likely to be very erratic compared with those of people in regular employment or dependent on benefits. This would mean that some might have very low recorded current income, putting them in the bottom tenth of the income distribution, but their expenditure is likely to be smoothed between their fluctuating periods of high and low incomes.

The picture of the *changes* in inequality is also different if an expenditure measure is used rather than an income measure. Although the distribution of expenditure widened over the 1980s, the very well-documented rise in income inequality was more rapid than the rise in expenditure inequality. All these differences are explored in much greater depth in Chapter 4.

The greater part of the empirical analysis presented here concentrates on income as the primary measure. In doing so, we follow the example of the great majority of authors who have worked on the income distribution. This reflects the availability and reliability of data as much as anything else. Nevertheless, there are a number of fundamental insights that can be gained by looking at expenditure as well. That is why the next two chapters, which deal with current income inequality and trends in income inequality, are followed by a chapter on expenditure inequality. The chapter on income dynamics (Chapter 9) also adds another dimension to the use of snapshots of income. Both these measures, and more, need to be considered in forming a comprehensive view of inequality. There is no single, correct way of measuring the standard of living.

THE MEASUREMENT CHOICES

Having discussed the relative merits of income and consumption as measures of living standards, we continue in this section by exploring some of the issues associated with their measurement. Neither income nor consumption is a simple concept. There is a wide range of choices we can make about what exactly it is we are looking for, and depending on these choices, there are different sorts of information on which we can draw.

Given the terms in which the discussion regarding income and spending was couched, the natural first question relates to the period of measurement. For example, should we choose to measure income (or spending) over a week, a year, a lifetime, or some other period altogether? One then needs to consider what should count in a measure of income or spending. One question is whether to measure income inclusive or net of taxes.

A further set of issues are raised when we stop thinking about an isolated individual and take account of the fact that most individuals live with others (their spouse, parents, children, or friends). Choices then have to be made over how this is taken into account. Are we interested in individuals, families, or households? How should households of different sizes be compared? These sets of issues are dealt with in the section below headed 'WHOSE INCOME?'.

The Time Period

Income and spending both represent flows of resources (as opposed to wealth, for example, which is a stock). If these flows vary over time, then the choice of time period matters.

An initial distinction can be drawn between measuring income or spending at a snapshot (over a relatively short period of time) and measuring them over a lifetime. The difference between comparing snapshots and comparing lifetime totals should be evident from the discussion about the choice of income or expenditure in the previous section. Let us return to our example individual who has periods in education, work, and retirement.

His weekly (indeed also his annual) income will vary—starting very low, rising as he enters work, and then falling again on retirement. Despite the predictions of the life-cycle hypothesis, his expenditure might well follow a similar (if flatter) pattern.[4] If we then compare two identical individuals at different points in their life cycles, we can see that comparing snapshots of income (or spending) will make the population appear unequal. Comparing lifetime amounts, on the other hand, will result in an appearance of equality.

Even within the period spent in work, people's earnings vary systematically with age. They tend to start off low, increase rapidly until age 40 or so (depending on a number of factors including skill levels), and then stabilise. Comparing the earnings of 20-year-olds with those of 40-year-olds might give the impression of significant inequality. But it is quite possible that the younger workers' earnings will rise such that when they reach 40 they will be the same as the older group's earnings were at 40. Lifetime earnings might be the same.

Of course, income and spending vary over much shorter periods than the lifetime or the working lifetime. As we show in Chapter 9 on dynamics, people move around the distribution from year to year for a whole host of reasons. People lose jobs and find jobs, are promoted, win the lottery. A well-paid director of a company could unexpectedly be made redundant and not find another job for several months. A builder might receive no income at all in some weeks, but in others receive enough to put much of it away for times when there is no income coming in. An actor or a musician might only have one big job a year, or several smaller ones scattered throughout the year.

As a general rule, measures taken over a longer time period will give a greater appearance of equality than will measures taken over shorter periods.

Which time period it is best to use will generally depend on what particular aspect of inequality it is that we want to measure. For the poor, a reduction in income for just one week could cause significant hardship. For others, snapshots will be misleading because they can smooth variations.

To avoid misleading impressions of inequality potentially created by looking at weekly incomes, some studies of living standards have chosen to focus on current *annual* rather than weekly income. This was the approach taken by the Royal Commission on the Distribution of Income and Wealth (see Layard, Piachaud, and Stewart (1978)),

---

[4] For example, spending is generally found to drop on retirement. See Banks, Blundell, and Tanner (1995).

whose evidence was also used in many other studies (for example, Atkinson (1983)). However, the data available on annual incomes in the UK are not as comprehensive as the data available on weekly income.

This lack of data is largely responsible for what drives us, in much of what follows, to use weekly incomes as our measure. But this is supplemented by the information on income dynamics in Chapter 9. This gives a good impression of how incomes do vary from year to year and how people move up and down the income distribution. And while we cannot hope to look at lifetime income profiles, we can look at how incomes differ within and between cohorts, and some attention is paid to this question in the same chapter. This, along with simple comparisons of income by age range, as in Chapter 2, can provide some indication of the degree to which the overall measures of inequality are influenced by life-cycle differences.

Expenditure

But can we not get around many of these problems just by looking at spending? As much of the discussion so far has emphasised, spending ought to be more evenly spread through the life cycle, and week-to-week or year-to-year fluctuations in income ought to be ironed out to some extent when it comes to looking at expenditure.

Sadly, there are two particularly important practical issues that make the issue of time period at least as important for spending as it is for income. The first relates to the difference between consumption and expenditure, and the second to the so-called 'lumpiness' of expenditure. They are related problems but are considered in turn.

Up to this point, the words 'consumption' and 'expenditure' have been used interchangeably. But they are not the same thing. People enjoy the consumption of many goods over a long period of time and not just at the time when the good was purchased. Cars, furniture, and washing machines yield a stream of consumption over several years. Very few goods are actually consumed instantaneously on purchase. All sorts of items, such as clothing, toys, and books, are consumed gradually after the expenditure on them. One could even argue that meals in restaurants and trips to the cinema or theatre—things whose consumption is not spread over time—are enjoyed in retrospect and therefore continue to convey well-being for some time after they have been bought.

When we try to measure a household's standard of living, it is the level of consumption, not expenditure, which is directly relevant. Current expenditure will *over*estimate consumption if someone has just made a large purchase of a durable item, but will *under*estimate the amount of consumption if they are making use of many goods for which they have already paid in the past.

Additional information about a household which could help rectify this measurement problem would be information about the access that households have to various consumer durables, such as televisions, cars, and washing machines. Even if income, not expenditure, is used to measure living standards, it would provide a fuller picture of the *command over resources* if consumption based on past expenditure (income) were included in the measure.

In principle, we could use information about a household's access to consumer durables to make an estimate of the value of the consumption of these goods. Added to current expenditure, this could provide a total consumption measure. In practice, this sort of estimation is very difficult. We would require not just 'yes' or 'no' answers about a household's access to durables, but also information about how much the goods cost and how long the household had been using them for. This sort of detailed information is not available.[5]

The second drawback with using expenditure is, as mentioned, its 'lumpiness'. On average, spending will be smoothed to take account of volatile incomes, but that does not mean that households will spend the same each week or even each year. This is in fact a problem closely related to the issue of the distinction between expenditure and consumption. To buy a consumer durable involves the outlay of a substantial sum of money at one time, with no repeat of that spending possibly for many years. It is this sense in which spending is often described as 'lumpy' in a way that *consumption* might not be.

As with income, most information on spending is available only over periods of a week or so (two weeks in the main data source, the Family Expenditure Survey). Ideally, information based on spending over longer time periods in combination with more information about ownership of durables would improve expenditure measurement and expenditure-based measurements of inequality. The fact that such

[5] There have been some attempts at measuring consumption from durables, such as Deaton and Muellbauer (1980).

information is not available is a serious problem, especially for interpreting single snapshots of expenditure inequality.

Nevertheless, two sorts of comparisons using such short-term expenditure data are of interest. The first type is the comparison between income and expenditure inequality. The second relates to changes in this relationship over time and in expenditure inequality over time. It is on these areas that Chapter 4 concentrates.

The Definition of Income

Income might seem to be a straightforward concept. But if we are interested in measuring living standards, then we may choose to adopt an income definition that is more comprehensive than the everyday idea of income. Obvious things to include in any income measure are earnings, incomes from various other sources such as investment and self-employment, and, as is discussed below, benefit income from the state. Beyond this, there are a whole host of things that could be included or excluded, depending on what aspects of people's resources or well-being we are interested in. In a comprehensive definition, one might also include the value of any capital gains that have been made. In this sense, one can think of income in any period as that amount of money that could have been spent without causing any reduction in real wealth levels.

There is no one 'true' income measure of living standards but many different ones. They range from the narrowest of measures—for example, cash income—to broader definitions that include everything from the value to individuals of company cars and other perks, to the value of government-provided education, health, prisons, and defence; they might be yet wider than this and include the value to people of the housework that they do or the leisure time that they enjoy.

Usually, some measure of disposable income is used for comparisons of living standards—that is, income that is available for spending on goods and services. For this reason, certain payments are generally deducted from income. Direct taxes and, in some cases, housing costs are examples of such deductions.

There are different sorts of questions one might want to ask about income. Depending on the question, a different definition or part of income might be appropriate. If one is interested in the impact of the state on incomes and inequality, one might want to disentangle *market* income from the effects of taxation and state benefits. Market income could be taken to include all earned income, investment incomes, and capital gains before taxes have been paid and benefits received.

Some of the more important issues are discussed below. They fall under three main headings—cash income, non-cash income, and deductions from income.

Cash Income

People receive income from a variety of sources, such as earnings, investments, pensions, and self-employment. It is obvious that these should be included in any definition of income. One other important source of cash income to households is the state. The government spent more than £90 billion on social security benefits in 1995–96 alone.

For some purposes, it may be informative to exclude social security benefits from income. For example, the distribution of pre-benefit income provides important information for the government when it sets means tests and benefit levels. If one wants to know how redistributive the social security system is, one can compare the distribution of pre-benefit income with the distribution of income that includes these payments. In terms of measuring living standards, however, it makes no sense to count income exclusive of all benefits. Many people are dependent on income from cash benefits from the government, and their living standards depend on this. In 1992–93, around 18 per cent of total household disposable income came from social security benefits alone.

The case for including one particular category of benefits is less clear-cut. How much some people receive in certain state benefits is a function of particular expenditures that they make. For example, Housing Benefit goes directly towards the payment of rent. If real rents were to rise, then someone on Housing Benefit would receive more benefits to cover their additional rent. Income including Housing Benefit would rise, regardless of the fact that this additional income would be immediately paid out again in rent, and disposable income after rent would remain unchanged.

If the rent rise reflected an improvement in housing quality, then it would seem correct to include as income the additional income to pay for it. On the other hand, if the quality of the housing remained unchanged, it would seem wrong to count the extra income as an addition to living standards. One way around this dilemma is to exclude any payments for housing costs from income altogether. This issue is discussed again below when we look at possible deductions from income.

The income measure that we generally use in this book includes all cash benefits, including Housing Benefit. A value is also assigned to certain benefits in kind, such as free school meals for children whose parents are in receipt of Income Support.

Non-Cash Income

Fringe Benefits and Government Benefits in Kind   As well as being paid a salary, many people receive 'perks', or 'fringe benefits', from their jobs. The most important example in the UK is the company car. Some agricultural workers are paid partly in the form of food or produce. These perks are not just presents from companies, but ways of paying employees more without giving them more money. If the companies did not give out these benefits, they would have to pay higher wages instead to attract and keep the same workers. The more important fringe benefits are in fact taxed as earnings. Otherwise, companies would have a huge incentive to offer more and more of these perks instead of money income, and avoid paying tax at all.

Since these sorts of perks really represent another form of earnings, the case is clear for including them in income wherever possible.[6] What of government benefits in kind?

Living standards are increased not just by government benefits, but by state provision of a whole variety of goods and services, notably education and health care. The level of such services will vary over time. Usage of them will also vary between individuals. Provision of such benefits in kind by the state, like the provision of benefits in kind by employers, increases the potential consumption of individuals. They can consume the health care, education, public parks, etc. without running down their stocks of wealth. On the other hand, there is no way in which they can be used to increase individuals' wealth.

There is a major question as to the reliability of any attempt to include the value of such services in income.[7] The prior question,

[6] No attempt is made to incorporate income from company cars and other such perks in the income measures used in this book. This is because of the difficulties in obtaining consistent information over time.

[7] There have been several studies that attempt to do exactly this; for example, CSO (1995) and Evandrou *et al.* (1993).

though, is would we want to include such estimates in any measure of income used to estimate inequality? Since many government benefits in kind are provided to people on the basis of need, it could be argued that they do not make those who receive them any better off than those who do not. The fact that the elderly make heavy use of medical services hardly makes them better off than a fit person who does not. People who have children cannot in any useful sense be said to be better off than those without because the state provides schools to which their children are legally obliged to go.

In considering how living standards alter over time or across countries, however, taking account of government spending might be more important. Higher levels of spending on better health care would make the sick of tomorrow better off than the sick of today. The same goes for education spending and children.

In this way, government spending is like a whole host of indicators associated with quality of life which might differ between individuals at any one time and across time. When considering reported changes in real incomes, it should be borne in mind.

Home Production and Leisure Time An issue closely related to the treatment of income and benefits in kind is how best to deal with the value of goods and services that people provide for themselves, most notably home production. Another issue is how to treat the value of leisure time.

Not all aspects of command over resources—which our basic definition of income should be measuring—are marketed. In the most simple case, farmers produce food for themselves. It is never sold, but it could have been. It has a value which ought to be assigned to the producer. Such issues of self-produced goods are of special importance in economies that are more agriculturally based than is the UK's, and it is particularly vital to include such production in income when international comparisons are being made.

Probably more important is the issue of services, such as housework or childcare, that are performed by household members. Such services can be bought in the market. In comparing the well-being of, for example, couples with children where both parents work in full-time paid employment and similar families where only one parent is in work, taking account of such issues is potentially important. Where both are in paid work, it is more likely that services such as childcare and cleaning will be paid for. Where one partner remains at home, he or she is likely to perform such services, rendering the household better off as a result. One way of valuing such services is to add the market value of such services to the income of the single-earner household.[8]

Taking account of this fuller definition of income certainly provides a basis on which to consider the way in which married couples are taxed in the UK. Each individual has a tax-free band of income or tax-free allowance. This means that if one member of a couple is in work and the other is not, they effectively only use one tax-free allowance. If they both work, they can each use their own allowance. So one earner earning £20000 pays more tax than two earning £10000 apiece. In so far as the single-earner couple enjoys a higher 'full income' than the two-earner couple, this might seem more appropriate than is immediately apparent if only cash income is considered.

One can take this type of argument one step further and assign some value to the leisure enjoyed by household members. This takes us further away from any defensible concept of income, but as a part of overall well-being, opportunity to enjoy leisure time must be of some value.

Comparing two individuals with the same income, we would normally say they are equally well off. But if one of them has to work twice as many hours as the other, it might be reasonable to argue that the person with shorter working hours was actually better off. If one takes this line of argument to its logical conclusion, then one might end up assigning relatively high living standards to the unemployed and others out of work.[9] In fact, this raises a number of difficult issues. Evidence on what makes people happy tends to show that actually having a job is at least as important as the amount of money that job yields (Clarke and Oswald, 1994). This contradicts the assumptions of simple economic models in which leisure is valued and considered a 'good' while work is a 'bad', but it corresponds with common sense. There is some range over which increased leisure is life enhancing (note the demands of many unions for shorter working weeks), but for many people the leisure gained from protracted spells of unemployment cannot be considered in the same way at all.

---

[8] See, for example, Jenkins and O'Leary (1994).
[9] This approach is taken in Pryke (1995).

Owner-Occupation   Ownership of the property in which one resides also confers a flow of benefits on the owner. It does not actually confer a flow of income but the effect is the same. Ownership means that rent does not have to be paid. If the property were not owned, the occupant would quite simply be worse off by the amount of the rent that would have to be paid to live in it. Looked at another way, if the owner were to move out and let the property to someone else, he would receive a flow of rent from the new occupier. Ideally, one would include a measure of imputed income from owner-occupation in a full income measure.

It is in fact rather hard to measure such income, and the debate in the UK has become focused in recent years on a related but slightly different issue—that is, whether income (excluding imputed income from owner-occupation) should be measured before or after housing costs are deducted. We move on to consider this next.

Deductions from Income

Income before or after Housing Costs?   To some extent, simply deducting housing costs—rent, mortgage, etc.—from income should give us a flavour of the income relativities that we would see from a measure of full income that included imputed income from owner-occupation. For example, outright owners would be seen to be better off than renters or mortgagers. On the other hand, such a measure does not distinguish between owners of valuable houses and owners of hovels.

The main advantage of deducting housing costs from income is that it can allow us to correct for one of the most important regional differences in living costs in the UK. Housing costs a great deal more in some areas than in others.

On the other hand, there are also strong arguments for keeping the cost of housing included in income. If people do *decide* to spend a large part of their income on buying a big house, then it must be contributing towards their living standard. And in any case, housing is not unique in that there are many other things that people have little choice over whether to purchase or not—for example, food and clothing—and the price of these can also fluctuate (though nothing like so much) from region to region regardless of differences in quality.

Whether incomes are measured before or after housing costs have been paid makes a considerable difference to the measured level of inequality and, to a lesser extent, how inequality has changed over time. Income measured after housing costs (AHC) is more unequally distributed than income before housing costs have been deducted (BHC). This reflects the fact that housing costs in general take up a higher proportion of the incomes of poorer households than of those of richer ones.

One potentially serious problem with the BHC measure is that it includes Housing Benefit as an income. This means that renters receiving HB appear to be better off than owner-occupiers not receiving it and as rents go up they might appear to be getting better off over time. But changes in the generosity of the Housing Benefit system would not be picked up by an income measure excluding it. Again, there is no obviously right answer. By making use of two measures, one before and one after housing costs, we can get around some of these problems.[10]

Income before or after Taxes?   The measure of income that is used in much of this book is measured net of personal direct taxes—that is, those that are levied directly on individuals. The main ones in the UK are income tax, National Insurance contributions,[11] and council tax. For the purposes of assessing people's living standards, what is relevant is income available to be spent on goods and services. The amount that an individual pays in personal direct taxes cannot be spent, and so cannot contribute to household living standards.

This deduction does have one serious implication. Because we do not impute the value of government-provided services in income, it appears that a rise in direct taxes unambiguously reduces the living standard of the person who has to pay the extra tax. What we do not do is measure the increase in living standards that would result from extra government spending, funded by tax increases, on health or education or law and order.

This treatment of direct taxes is different from the way we treat other taxes in the measurement of living standards. In fact, only about

[10] See Johnson and Webb (1992) and Harris and Davies (1994) for a detailed discussion of the role of housing costs and Housing Benefit in measuring income.

[11] Only employ*ee* National Insurance contributions are deducted. Ideally, we would like to deduct employ*er* NICs as well, since (despite their name) the incidence of these taxes is also on the individuals employed rather than the firm.

one-third of all tax revenue collected in the UK comes from direct taxes. The remaining two-thirds of tax revenue comes from indirect taxes and other taxes that do not immediately impinge on individuals (most notably taxes on companies). Indirect taxes are those that are raised from the sale of goods. The main indirect taxes in the UK are value added tax (VAT) and the excise duties on alcohol, tobacco, and petrol. In the other group are taxes such as corporation tax and employer National Insurance contributions. Although these are levied initially on companies, in the end it is individuals who will bear the cost of them. Neither of these sorts of taxes are deducted from incomes in the way that the personal direct taxes are.

The question of the role of indirect taxes is more difficult than that of the role of direct taxes. Although they are not deducted directly from incomes, a rise in indirect taxes clearly makes people worse off (subject to the same caveat about public spending out of additional tax revenues). One could argue that they are voluntary in the sense that paying for an increase in the tax on petrol could be avoided by reducing petrol consumption. This does not change the fact that the tax makes people worse off—they would clearly have been better off without the tax. If they bought petrol before the tax increase, they did so because they maximised their welfare by so doing. Their welfare is less afterwards if they no longer buy it because of the tax increase. In any case, there is an analogy with direct taxes when their payment can be reduced by reducing labour supply and thus earning less.

Now suppose there were a proportional indirect tax system so that indirect tax was levied at the same percentage rate on all spending. Furthermore, suppose all income were spent. In this case, it clearly makes no difference to inequality whether we use income net or gross of indirect tax. In comparing living standards over periods during which the rate of indirect tax changes, however, it might appear that the choice would matter—an increase in the rate of indirect tax would just make people worse off. But this is in fact just a question of using the appropriate price deflator. Living standards net of indirect tax should be compared using a price deflator that is also net of indirect tax. If indirect tax is not netted off income, then neither should it be netted off the price index. In fact, the actual retail price index used in the UK is appropriate for comparing incomes over time that are not adjusted for indirect taxes.

To extend the situation, suppose some people now start saving money. In this case, the savers pay less tax than the spenders. If people expected the indirect tax system to continue in its current form, its existence should not make any difference to choice over whether or not to save; savings have to be used eventually to spend. So there is no real sense in which the savers are better off than the spenders, despite the fact that in the current period they are paying less in indirect tax. So netting indirect taxes off incomes leads to a false differentiation between savers and spenders.

The problem occurs when, as is in fact the case, indirect taxes are not uniform and do not rise uniformly. In this case, individuals' relative purchasing powers can be altered by indirect taxes and so indirect tax changes do have differential effects on welfare. Over time, they will impact on measures of inequality. The overall effect on average living standards could be factored out by comparing real living standards based on the retail price index, but the distribution might be changed. Similar arguments apply to other price changes. It is quite possible that the prices of some goods enjoyed disproportionately by a part of the population will go up faster than others. If the price of children's clothing were to rise especially quickly, then families with children might be left worse off. So looking at the effects of indirect tax changes over time would be a part of constructing appropriate price indices for all the various parts of the population.

In fact, Crawford (1994) has shown in his comprehensive study that price changes induced either by indirect tax changes or otherwise have made remarkably little difference to relative real living standards. In other words, the price inflation faced by various family types and parts of the income distribution has been similar over prolonged periods. The major exception to this rule has been the exceptionally fast growth in the price of rented housing, which has impinged particularly severely on some groups in the lower part of the income distribution. These findings are of course contingent. They need not apply over any other periods or any other populations than those to which they were applied. But they should inform a good deal of what follows. For much of our work on the income distribution over time, we need not concern ourselves with differential price indices and therefore with changes in the indirect tax system. The exceptional case of housing has already been considered separately.

There remains a case for netting off indirect taxes when comparing individuals at a point in time to take account of government policy that affects some groups more than others. There is also a case against because to do so would wrongly distinguish between those who are spending and those who are saving. And if we were to accept the case for netting off indirect taxes, there would also be a strong case for taking some account of taxes such as corporation tax whose immediate incidence is on companies. Companies pay corporation tax, but companies are just legal entities. Shareholders and customers will each bear part of the burden of such taxes.[12]

In what follows, only direct taxes are netted off incomes when looking at living standards. This appears to be the best way of getting at relative living standards based on income as providing a potential level of consumption. The discussion of the past few paragraphs should be borne in mind when interpreting all that follows.

WHOSE INCOME?

Up to now, we have been talking about individuals as though they lived alone. But in fact they often live in households with other people. Much discussion of the income distribution is in fact based on the household as the basic unit of analysis with the income of the household assumed to be shared equally among its members.

The most obvious problem with the opposite strategy of measuring incomes received by individuals as though they lived alone would relate to the treatment of children. That they do not receive any income in their own right is not the most relevant fact when it comes to measuring their living standards. In general, they share in the living standards of their parents.

Another more relevant question relates to the relationship between adults within a household. Again, it is easy to see why one would not generally be interested in simply comparing individuals' receipt of incomes. Many married couples have one spouse earning and the other not in paid work. It would be misleading to treat a married couple where one partner is earning £25 000 per year and the other nothing as two separate income units—one well off and the other poverty stricken.

There tends to be *some* sharing within households. The trouble is

---

[12] Dilnot, Kay, and Keen (1989) provide a framework in which the incidence of different taxes can be considered.

that this fact tells us nothing about how much sharing there is. There is certainly no reason to suppose that the sharing is everywhere *equal*.

Some economic theories have been developed to predict how sharing within a household or family unit is determined. For example, Becker (1981) sets out a model in which an altruistic 'head' of the family makes net transfers to other family members. This model could provide an underpinning to the assumption that there is equal sharing of resources within the family. But the model simply assumes altruism of the head—or main income recipient—which is precisely the assumption of complete sharing. Alternatively, Sen (1984) sets out a model of 'co-operative conflict' in which household members co-operate with others in order to make the pool of resources larger, but conflict over how to divide up these resources. This could underpin a number of different sharing scenarios, depending upon the relative power of the different household members. Here it is the relative power of the household members that matters.

The problem remains, though, of distinguishing between the validity of the different models. Reality is that in some households there will be full sharing, in others there will be varying degrees of sharing. Identifying how much there is in any given household can be attempted in two sorts of ways. The first relies on asking household members how budgeting occurs within the household and who has control over resources. The extent to which such questions can elicit correct answers is debatable. Datasets with good-quality income information rarely have good-quality information on sharing rules.

A second methodology might be based on observing consumption behaviour among similar households with differing income shares. It might be possible to deduce something by looking at the sorts of goods bought by households in which all the money is earned by the husband and the sorts of goods bought by households where half or more is earned by the wife. This apparently straightforward procedure is, however, complicated by formidable problems of estimation. Having two workers in a household will require different sorts of spending—on transport to work and labour-saving domestic appliances, for example. More serious is the fact that two-earner households are likely to be composed of different sorts of people with different sorts of preferences from those of single-earner households. So differences in consumption behaviour might simply reflect disparities in preferences between the different sorts of household.

The actual extent to which there is sharing between household members remains an open question. It is an important one because inequality measured on the assumption that there is full sharing will be lower than that measured on alternative assumptions. Measured changes in inequality over time will also be affected. Suppose the degree of sharing increases over time, either because of social changes or because receipt of income within the household becomes more equal.[13] In such cases, an increase in overall inequality measured on the full-sharing assumption might be rather greater than the actual increase would have been if the degree of sharing had been accurately measured.

We have approached this problem so far as a question relating to sharing between husbands, wives, and their dependent children. Such a grouping is more accurately referred to as a family or benefit unit than as a household. Households can contain several benefit units. Most commonly, this might involve parents living with their grown-up children who are no longer dependants, or people living with elderly relatives. One also finds households containing a number of unrelated individuals. As there is within families, there is often sharing of incomes between families within households. This might not be explicit or complete but will generally at least involve access to the same consumer durables, heating, food, and other common living arrangements.

In much of what follows, we accept the usual practice of assuming complete sharing within households. This is a controversial assumption. As Lazear and Michael (1988) point out, the equal-sharing assumption is one that is generally made because the data available to us are so limited, not because it is an accurate reflection of what actually goes on: 'we know of few clear statements which defend the use of the family or the household as the appropriate, rather than the convenient unit of analysis'.

A much fuller discussion of this important area is available in Jenkins (1994a).

---

[13] Webb (1993) shows that income receipt within households did indeed become more equal over the 1970s and 1980s, largely as a result of increased labour market participation among married women.

Comparing Households of Different Needs

Where any unit other than the individual is taken as the basis for comparison, it becomes necessary to make some adjustment to income to take account of the different sizes of the income units. Married couples need more money than single people, but not twice as much. This is another way of expressing the old adage 'two can live as cheaply as one'. In a similar way, families with children need more money than those without.

Even this, though, is a debatable point. One of the problems associated with measures of the welfare costs of children is that, for the adults concerned, there is presumably some welfare gain associated with the simple existence of the children. To put this another way, people, by and large, choose to have children—they are not delivered unexpectedly by storks or found under gooseberry bushes. That being so, it might seem odd to argue that the parents then need more money to return them to the level of welfare they attained before having the child. Partly because such welfare gains from children cannot usually be measured, it has been argued that it is not possible for economists to identify welfare differences between families with and without children.[14]

This is an important and complex issue but not one into which we delve any deeper here. Rather, we simply follow the generally accepted convention that one should take account of the existence of children as well as of second and subsequent adults in a household, when comparing incomes. For it is considered that where there are children to provide for, more money is needed to reach a particular standard of living.

Comparisons between families or households of different sizes are generally made through the use of *equivalence scales*. An equivalence scale is defined as the ratio of the cost (to a household) of achieving some particular standard of living, given its demographic composition, to the cost of a 'reference' household achieving that same standard of living. So if the reference household is taken as a childless couple and the equivalence scale for a couple with children is estimated at 1.5, this implies that a couple with children need one-and-a-half times as much income as a childless couple to reach a particular standard of living.

---

[14] See Pollak and Wales (1979).

Estimating Equivalence Scales

One can identify at least three broad types of methods for estimating equivalence scales. The first can be characterised as needs-based on nutritional requirements, the second as welfare-proxy-based, and the third as utility-based.

Nutritional or Needs-Based Scales   Nutritional or needs-based scales developed from work on poverty measurement which required ways of estimating who fell below certain absolute poverty lines. To do this, subsistence minimum incomes had to be calculated for families of different types. Such incomes had clearly had equivalence relativities built into them. Rowntree (1901) was one of the most famous users of such scales in his studies of poverty in the city of York. In setting benefit scales, Beveridge used information on minimum necessary incomes for different family types. This type of approach has been taken up in recent years in the work of the Family Budget Unit at the University of York.[15]

Welfare Proxy Methods   The traditional method used by economists for estimating equivalence scales has essentially sought a proxy for welfare—a household characteristic that has an indirect relationship to the welfare of the household is used and compared across households.

The oldest method of attributing welfare levels to differing consumption bundles stems from 'Engel's Law'—the famous observation that as households become wealthier, the proportion of their expenditure that is spent on food decreases. Food is a necessity. If you only have a few pounds a week, you are likely to have to spend most of your money on necessities such as food. If you are rich, then food will take up a smaller share of your budget. Consequently, Engel (1895) postulated that the budget share of food would be an appropriate indirect measure of welfare.

Such a negative relationship between food share and income is illustrated for two hypothetical household types in Fig. 1.1. The presence of a child in a household will, other things being equal, increase the share of expenditure spent on food (from W_0 to W_1, say, at income Y_0), implying that welfare has fallen (under the Engel

---

[15] See, for example, Bradshaw (1993).

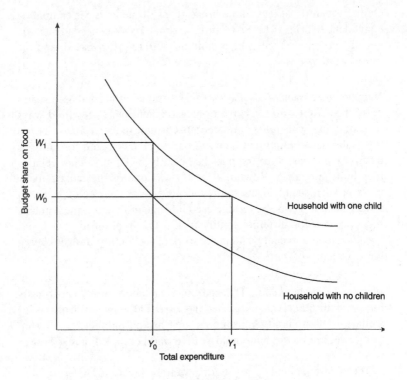

Fig. 1.1. Engel curves and equivalence scales

identifying assumption). The size of the fall in welfare can be measured in this situation by the amount of income needed to restore that food share to its original value on the new Engel curve—that is, the difference between Y_1 and Y_0 in Fig. 1.1.[16]

A more general method than just using food share is to use a share of necessities as a whole including, for example, food, fuel, and housing costs. The argument, of course, runs in exactly the same

---

[16] The consistency of the negative relationship between the food expenditure share and total expenditure is well documented for the UK (see Banks, Blundell, and Lewbel (1992), for example). Some simple Engel scales are estimated in Banks and Johnson (1993).

way and the problems are the same. The basic problem is just that we have no way of proving that the share of food or necessities in a household's budget is an appropriate proxy for welfare.

Another commonly used methodology has been the Rothbarth (1943) method. Instead of taking food share as an indication of welfare, consumption of 'adult goods' is used to indicate welfare. Two households are considered equally well off if they consume the same quantity of such goods (alcohol and adults' clothing might be good examples). Obvious problems with this method are associated with the fact that people's preferences for goods such as alcohol are likely to change when they have children.

Utility-Based Methods   The 'reduced form' approaches discussed so far obviously require some rather strong identifying restrictions. Why should we believe that food share is an accurate representation of welfare? The assumption that preferences for 'adult goods' remain unchanged when a child enters the household is possibly an even stronger one. Utility-based estimation of equivalence scales developed as a response to these sorts of criticism and seeks to model the complete system of household economic preferences—allowing the presence of children to enter into preferences in the way that best explains observed economic behaviour. Having calculated demand responses, economic theory can then be used to calculate the implied welfare levels for households of differing demographic characteristics with other factors held constant.

This methodology, though, requires its own sets of assumptions, for there are many different sets of preferences over demographic composition that will generate the same observed demand behaviour, and each will generate a different equivalence scale. In other words, some sort of identifying assumption has to be 'smuggled in' in order to end up with an equivalence scale. The main difference between this and the previous methods is that the assumption can be said to be slightly less '*ad hoc*' and to be based in economic theory. The danger is that this will lead to the estimated scale appearing to have an objective validity that it does not in fact possess.

It is also worth mentioning one other aspect of equivalence scale estimation at this point. That is simply that consumption decisions are not made at one point in time in a manner that is completely independent of previous and future decisions. One might reasonably expect families who are intending to have children to save in response to that expectation. In that case, the costs of children will be borne not just while they are observed in the household but prior to that as saving occurs and possibly later when debts are repaid. If so, equivalence scales estimated just on the basis of snapshots will be misleading.[17]

Just a very few of the huge number of estimated equivalence scales are illustrated in Table 1.1.

Effects of Different Scales

Given the wide range of available scales, it is not surprising that different studies of the income distribution use different scales. The use of different scales can result in quite different measures of inequality and rankings between different population groups, and there is an extensive literature on this subject. Much of the effect of different equivalence scales in relation to the UK was summed up in papers by Coulter, Cowell, and Jenkins (1992) and Banks and Johnson (1994).

One can think of comparing different scales by considering the weights given to extra individuals in a household. At one extreme, each individual is given full weighting equivalent to the weight of the first adult in any household. In this case, we are looking at a per capita measure of income—household equivalent income is just household income divided by number of people. At the other extreme is the case where no weight at all is assigned to second and subsequent individuals. In this case, we just compare raw household incomes. Of course, all equivalence scales lie in the range between these extremes.

The trouble with making no adjustment for household size is that big households will appear to be better off than they actually are. To get round this problem, equivalence scales give weight to extra household members. This will initially have the effect of reducing inequality as the incomes of the richer, larger households are gradually reduced by dividing their cash incomes through by progressively larger scales. At some point, this process is likely to come to an end as the equivalent incomes of the large households start to fall below those of the small households. Thus there tends to be a U-

[17] Further discussion of this issue is contained in Banks, Blundell, and Preston (1991 and 1994).

shaped relationship between recorded inequality and the weight given to extra household members.

In the UK in recent years, differentiating between the effect of additional adults and the effect of additional children has been shown to be important. Families with children do not tend to have much more income than those without. As a result, increasing the weight given to children in any equivalence scale, from even a very low base, tends to result in greater measured inequality as the equivalent incomes of families with children fall further and further behind those of childless families. Essentially, the more weight an equivalence scale gives to children, the greater the inequality that will be recorded if that equivalence scale is used.[18]

Finally in this section, a word about the particular equivalence scales we use in most of what follows. They are the scales derived by McClements (1977) and used extensively by, for example, the Department of Social Security (DSS) and the Central Statistical Office (CSO). The McClements scales for two income measures, before and after housing costs, are presented in Table 1.2. A childless couple is taken as the base, and so the scale for a childless couple is set at one. For all other family types, their income is divided by the appropriate scale to bring it into equivalence with that of a childless couple. So, for example, a single person would have their income before housing costs divided by 0.61, a couple with two children aged between 8 and 10 would have their income divided by 1.46, and so on.

DATA

Because it is never possible to know enough about everybody in the population, any description of the income (or expenditure) distribution must rely on data showing the incomes and other circumstances of a sample of people drawn from the population. No empirical analysis can get off the ground without such data.

There are no datasets that are perfect. The only one that is based on the entire population is the census. The costs of obtaining full compliance in the census are high—it is conducted only infrequently

---

[18] But do note that this is an entirely data-dependent finding. It is not a universal rule of equivalence scales applicable at all times and in all places.

Table 1.1. Some equivalence scales

Author	Date	Country	Group (couple + . . .)	Scale	Comments
Needs-based scales					
Engel, 1895	1895	Belgium	Child 0–5	1.19	
			Child 6–14	1.31	
			Child 15–18	1.41	
Rowntree, 1901	1901	UK	One child	1.24	Based on dietary needs plus 'minimum necessary expenditure' in 1899 survey of York.
			Two children	1.61	
			Four children	2.22	
Muellbauer, 1980	1979	UK	Child 0–5	1.26	Based on costs of consuming recommended daily calorie intakes.
			Child 6–18	1.45	
Welfare proxy scales					
Muellbauer, 1979	1968–73	UK	One child	1.17	
			Two children	1.34	
			Three children	1.51	
Muellbauer, 1979	1968–73	UK	Child 0–5	1.08	
			Child 6+	1.21	
Ray, 1986	1968–79	UK	Child 0–2	1.01	Linear Engel curves.
			Child 3+	1.23	
Ray, 1986	1968–79	UK	Child 0–2	1.00	Non-linear Engel curves.
			Child 3–5	1.03	
			Child 6–18	1.11	

Nicholson, 1949	1937–38	UK	Child 0–4 Child 5+	1.11 1.16	Rothbarth estimation; 'medium' welfare (60–70 pence weekly expenditure).
Utility-based scales					
Ray, 1986	1968–79	UK	One child Two children Three children	1.21 1.42 1.63	
Blundell and Lewbel, 1991	1970–84	UK	Child 0–2 Child 3–5 Child 6–10 Child 11+	1.09 1.14 1.16 1.18	
Banks, Blundell, and Preston, 1991	1970–88	UK	One child Two children Three children	1.22 1.50 1.84	Life-cycle scale—not strictly comparable

Table 1.2. McClements equivalence scales

Household member	Before housing costs	After housing costs
First adult (head)	0.61	0.55
Spouse of head	0.39	0.45
Other second adult	0.46	0.45
Third adult	0.42	0.45
Subsequent adult	0.36	0.40
Each dependant aged:		
0–1	0.09	0.07
2–4	0.18	0.18
5–7	0.21	0.21
8–10	0.23	0.23
11–12	0.25	0.26
13–15	0.27	0.28
16+	0.36	0.38

Source: McClements, 1977.

(every ten years), and asks for just a limited amount of information from each respondent.

All the other datasets that are available are drawn from a small sample of the population. To the extent that the chosen sample is representative of the whole population, this is not a problem. But even if a truly representative sample could be found, because most surveys are voluntary those replying to the survey will not be representative. Typically, for example, the very rich and the frail are less likely to reply than are others. Where there is compulsion, for example in tax-payment-based data, coverage is only of a part of the population.

Some of the problems created by data in which certain groups are under- or over-represented can be overcome by using differential *grossing* factors. Households of a type that are under-represented are taken to represent more households in the real world, and therefore given a greater weight, than are those that are over-represented. Where the over-/under-representation is in several different dimensions—age, employment status, and region, for example—this process can be very complex.[19] More usually, one dimension is chosen which means that the data represent the population correctly in that dimension but not in others. For example, it might be adjusted to

[19] See, for example, Atkinson, Gomulka, and Sutherland (1988).

represent the correct number of single people, couples, and pensioners but this might still leave problems with the relativities between the numbers in and out of work.

Even among those who are interviewed, there is no guarantee that what they report about their incomes and expenditures will be fully true. Most people who smoke or drink underestimate the amount they spend on alcohol and tobacco, for example. Many are also loathe to give information about investment income and capital holdings even where they have no compunction about providing details about the rest of their incomes. It is also easy for people to be confused about some of their income receipt and to provide wrong information as a result of genuine errors. This tends to be particularly problematic in distinguishing between certain sorts of social security benefits.

Indeed, as we discuss in Chapter 7 on taxes, where taxes and benefits are concerned one has the option of modelling payments and receipts as opposed to just taking as given what people report. The reported amounts are often not consistent with reported incomes and other circumstances. Whether it is preferable to take reported amounts or to model them will generally depend on the precise question we are trying to answer.

A separate group about whom it is worth making specific comment are the self-employed. Certainly in the data we use here, and as far as we can ascertain in most datasets, there are serious problems about interpreting the recorded information as it pertains to this group. The relationship between their recorded profits—their net income—and their apparent living standards as measured by their spending tends to be obscure. Many of the self-employed recording small or even negative profits actually appear to have rather high levels of spending. These issues are discussed in more depth in Chapter 4.

Different Datasets

Most of the data that have been used in describing the distribution of income, particularly in the UK context, have been cross-sectional— that is, they are based on a different sample of individuals each year. This allows one to measure the distribution of income at any one time and to see what is happening to the overall distribution over time. What it does not allow one to do is see what happens to individuals over time. For this one needs *panel* data—one needs to re-interview the same people year after year to see what happens to them.

The main source of data used in this volume is a cross-sectional dataset called the *Family Expenditure Survey* (FES) which is an annual survey of around 7000 households drawn from the UK population.[20] It is collected for the government principally to provide information on expenditures for the construction of the retail price index. But it also contains extremely detailed information on incomes. This combination of data on incomes and expenditure for more than thirty years is central to much of the empirical work in the following chapters.[21] As explained above, it is necessary to apply grossing factors to these data, and this is done by ensuring that different types of family are correctly represented such that the grossed-up data contain the same number of families with children, single pensioners, and so on as the population as a whole. A separate grossing factor is given for each family unit[22] within each household, depending upon to which of seventeen different categories it belongs. The categories are determined by age, sex, marital status, and number of children.

One further adjustment is made to the data to take account of the undersampling (and variable undersampling) of the rich. This problem is not dealt with by usual grossing methods as rich families of all types are undersampled. Instead, extra information from the *Survey of Personal Incomes*, a dataset based on income tax records, is used to ensure that the dataset eventually used is fully representative even of the very richest.[23]

The other main dataset that we use to supplement this is a panel survey called the *British Household Panel Survey* which forms the basis of the analysis in Chapter 9. This survey re-interviews people from the same 5000 households each year. This is a relatively new survey and at the time of writing there are only three years of available data. Nevertheless, the results already arrived at from the data add a great deal to our knowledge, and further work on new waves of

---

[20] Being a household survey, it omits, and therefore so do our analyses, information on the non-household population including the homeless, prisoners, and those in residential or nursing homes. For an analysis of the possible effects of excluding these groups, see Evans (1995).

[21] There are a number of publications devoted to an examination of the reliability of income data in the FES, including Kemsley, Redpath, and Holmes (1980), Atkinson and Micklewright (1983), and Jones, Stark, and Webb (1991).

[22] As will be explained, a family unit (or more precisely benefit unit) is very close to the nuclear family, being made up of an adult, any spouse (including cohabitees), and any children under 16, or under 19 and still in full-time education other than higher education.   [23] Details of this procedure are given in Goodman and Webb (1994).

these data will be invaluable in extending our knowledge of the dynamics of the income distribution.

THE MEASUREMENT CHOICES: INEQUALITY

All that has been discussed so far in this chapter has been focused on the issue of how the welfare or income or expenditure of any individual or single household should be measured. Nothing has yet been said about how these should be compared, or how to arrive at a measure of inequality. In this section, we present a very brief discussion of this issue directed mainly at explaining the inequality measures used in the rest of the book. This is not intended to be any sort of a general discussion of how inequality should be measured or what the properties are of the innumerable measures of inequality that have been developed by economists and statisticians. For a comprehensive survey, the interested reader is referred to Morris and Preston (1986) and Cowell (1995).

Over a number of years, summary statistics can be extremely valuable in giving quick impressions of movements in inequality— did it go up or down, and by how much? By decomposing certain measures, it is also possible to see what incomes and what groups of the population are having the greatest effect on inequality levels and movements. In looking at individual years, though, much more detail than can be gleaned from any such statistic is needed before much can be learned about the degree and nature of inequality.

A summary inequality measure is just that—a measure that can summarise the degree of inequality in one number.

In this section, we set out the measures of inequality that are used in the rest of the book. All of these measures have some things in common. For example, if everybody receives a 10 per cent increment to their incomes, then the degree of inequality should not change. Another way of putting this same point is that inequality measures must not be affected if all incomes are re-scaled by the same number; for example, 'if incomes are measured in pounds sterling or pence' (Atkinson, 1983). Another thing that they have in common is that if two identical populations are combined, then the inequality measured in the combined economy should be identical to that in the separate ones.

But each measure of inequality is slightly different. Not all of the summary measures of inequality that we use in this book would answer a very basic question in the same way—namely, 'which of two distributions of income is more unequal?'. There are some measures of inequality that would put more weight on the gap between the incomes of the very poorest and those of the richest. Others might place less weight on this and find it more important that the bulk of the population near the middle of the distribution have incomes that are relatively near those at the top.

The Gini Coefficient

The most commonly used summary measure of inequality, and one that we use extensively, is the *Gini coefficient*. The derivation of the Gini coefficient is most easily understood using a particular graphical presentation of the income distribution known as the *Lorenz curve*. The Lorenz curve summarises the distribution of incomes (or expenditures) by plotting income shares against population shares. The entire population is lined up in order of income along the horizontal axis, with the poorest in the left-hand corner and the richest in the right. Along the vertical axis is the *cumulative income share*—that is, the share of the total income going to each proportion of the population.

A Lorenz curve is drawn in Fig. 1.2. But before considering the curve, consider the 45-degree line that joins the origin to the top right-hand corner of the diagram. This could itself be a Lorenz curve representing an economy where everyone receives the same income, and so there is complete income equality. The bottom 1 per cent of this population gets 1 per cent of the income, the bottom 20 per cent gets 20 per cent of the income, and so on. At the other extreme, if one person received all the income and everybody else received none, the Lorenz curve would be made up of the horizontal axis and the right-hand side vertical axis. This is characterised as a situation of complete inequality, even though it should be noted that everyone except one person receives the same income (zero). This is because the Gini coefficient that is derived from this Lorenz curve reflects how much the income that poorer individuals receive differs relative to the income accruing to richer individuals.

The Lorenz curve for a typical income distribution, on the other hand, looks like the curved line in Fig. 1.2. Here the curve bulges out from the 45-degree line, since those with the lowest incomes take up a higher share of the population than they do of the total income. For

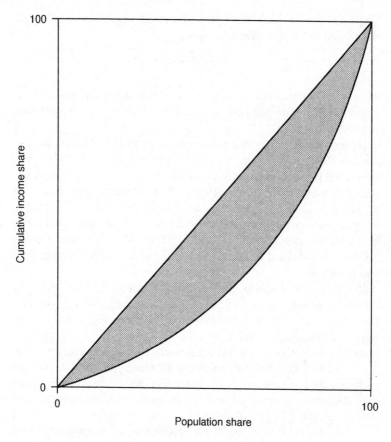

Fig. 1.2.   A typical Lorenz curve

example, in 1992–93 in the UK the bottom 10 per cent of the population received just under 3 per cent of total income. The bottom 90 per cent received just less than 74 per cent (meaning that the top tenth took up more than a quarter of the entire income of the population). Income shares such as these are a popular way of describing how concentrated income is in the hands of a few.

The Gini coefficient is simply the ratio of the area between the Lorenz curve and the diagonal (the shaded area on Fig. 1.2) and the whole area under the diagonal. It is just a standard and convenient way of summarising the degree to which incomes are concentrated.

There have been many ways proposed of displaying the Gini algebraically. One such is shown below:

$$I_{Gini} = \frac{2}{n^2\bar{y}} \sum_{i=1}^{n} i(y_i - \bar{y})$$

where n is the number of income units, $\bar{y}$ is mean income, i is the position in the distribution, and y_i are individual incomes ranked from poorest to richest.

If the Lorenz curve that represents a particular distribution lies entirely inside another one, then so long as it is agreed that a transfer of income from a richer person to a poorer person reduces inequality (see Atkinson (1970) and Jenkins (1991)), it can unequivocally be said that the economy represented by the outside Lorenz curve is more unequal than the one represented by the one that lies inside. The ratio of the shaded area to the whole area under the 45-degree line is larger for the more unequal economy, and so the Gini coefficient rises with rising inequality.

But it is also possible that the Lorenz curves for two different distributions may cross. An example of two hypothetical distributions where the Lorenz curves cross is shown in Fig. 1.3. In economy A, represented by the Lorenz curve marked A, the bottom two-fifths of the population have a greater share of the total income than do the bottom two-fifths in economy B, represented by the Lorenz curve marked B. But then curve A cuts curve B from above, and over the next fifty percentiles of the distribution it lies outside curve B, intersecting it again at around the ninetieth percentile. This can be taken to mean that in A the poorest have a relatively high proportion of the total income and those in the middle of the distribution have a relatively small amount, while those at the top in A have more than those at the top in B. The fact that the Lorenz curves cross stops us making definitive comparisons over the degree of inequality in the two situations. But a knowledge of their shapes and points of crossing tells us a great deal about the relative shapes of the distributions.

If Lorenz curves do cross, then the way in which different inequality measures rank two different distributions depends on the importance each gives to inequality at different parts of the distribution. Atkinson (1970) illustrates the pervasiveness of the problem of ambiguity in the ranking (in terms of inequality) of different income distributions by looking at the Lorenz curves for twelve different

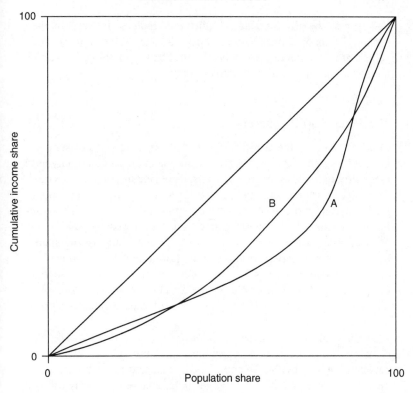

Fig. 1.3.   Crossing Lorenz curves

countries, based on income data collected by Kuznets (1963). He found that out of twenty-six possible pairwise comparisons, ten showed Lorenz curves that intersected. Jenkins (1991) used data on the UK income distributions over the 1960s, 1970s, and 1980s to make 171 pairwise Lorenz curve comparisons. He found that in about 30 per cent of cases the Lorenz curves crossed.

There is no single objective way to decide, in the case where Lorenz curves cross, which of two distributions is more unequal. Different inequality measures may rank the inequality in distributions differently, depending on several different features or 'properties' of the measure chosen. This should not necessarily be a worry, however. As Morris and Preston (1986) point out, 'provided that one bears in mind the differing nuances of the different indices their empirical values may be illuminating'. If different indices do vary in their ranking of income distributions, then, so long as the indices themselves are understood, we might well learn something interesting about the distributions so described.

The Mean Log Deviation

Measures such as the Gini coefficient are useful as ways of summarising the degree of inequality and are among the most commonly used inequality indices. In this book, use is also made of two other inequality indices with the specific purpose of allowing the components of inequality and more especially the components of change in inequality to be identified. These indices are decomposable and allow answers to questions such as 'how much is contributed to inequality by different population groups?' and 'how much is contributed by different income sources?'.[24] We make particular use of them in Chapters 5 and 6 of this book where an attempt is made to explain the changes in the income distribution over the past thirty years on a consistent basis.

The first of these measures is the *mean log deviation* (MLD).[25] As its name suggests, it is a logarithmic measure of the relationship between each individual income and mean income. Like the other measures presented, it is relative invariant—equal proportionate changes in income leave the measure constant. It is also a measure that gives greatest weight to changes in incomes at the bottom of the distribution. Unlike the other measures so far looked at, it is not bounded by zero and one at the maximum, rather by zero and infinity. It is set out algebraically below:

$$I_{MLD} = \frac{1}{n} \sum_{i=1}^{n} \log \frac{\bar{y}}{y_i} \cdot$$

From our point of view, the value of the MLD lies in the fact that it is additively decomposable and can be used to show the breakdown of inequality within and between population groups. It can be rewritten as follows:

---

[24] Jenkins (1995) makes use of similar decompositions to look at inequality in four years between 1971 and 1986. Mookherjee and Shorrocks (1982) and Shorrocks (1982a and 1982b) are largely responsible for having developed the theory and decompositions.

[25] See Theil (1967 and 1972).

$$I_{MLD} = \Sigma_k \frac{n_k}{n} I_k + \Sigma_k \frac{n_k}{n} \log \frac{\bar{y}}{\bar{y}_k} \; .$$

where I_k is inequality within group k, $\bar{y}_k$ is group k's mean income, and n_k is the number of people in group k. The first term of the equation describes the within-group component of inequality and the second the between-group component. In other words, the first term describes the total amount of inequality within the k groups into which the population has been split. Naturally, one can recover from this the inequality within each individual group as well as the total level of within-group inequality. The second term essentially tells us how much inequality there would be if each member of each population sub-group had the mean income for that group so that the only inequality resulted from differences in the mean incomes of each group and there was no inequality within each group.

Half the Squared Coefficient of Variation

The second decomposable index that we employ is used in order to look at the effect of different types of income on overall inequality. The index is a very simple one, namely *half the squared coefficient of variation*:

$$I_{cv} = \frac{\sigma^2}{2\bar{y}^2}$$

which is just a measure of the variance normalised by the square of the mean. The variance is itself a simple inequality measure being a measure of the average distance between any one income and the mean income, but it varies with the mean. Half the squared coefficient of variation, being normalised by the square of the mean, does not suffer from this problem. It is, though, very sensitive to outliers (very high or very low incomes a long way from the mean).

The first step in decomposing this index is to show it as

$$I_{cv} = \frac{1}{2}(C_k^A + C_k^B)$$

where

$$C_k^A = \frac{\sigma_k^2}{\bar{y}^2}$$

and

$$C_k^B = \frac{\sigma_k^2 + 2\text{cov}\,(Y_k,\ Y - Y_k)}{\bar{y}^2}$$

where σ_k is the variance within group k, Y_k is income from source k, and Y is total income. The first term, C_k^A, represents the coefficient of variation of each income source k. It is clearly always positive and reflects the total coefficient of variation that there would be if all other sources of income were distributed equally. It is generated by replacing each other income source by its mean for each household, and hence is weighted by the overall importance of that income source to household income. The second term, C_k^B, represents the change in the total coefficient of variation that would occur were one source of income distributed equally. This can be negative where an egalitarian redistribution of one source would actually increase overall inequality, as would happen, for example, if a means-tested social security benefit were to be given an egalitarian redistribution.

This measure of inequality is used extensively in Chapter 5 where we consider the role of different sources of income in contributing to overall inequality. There, we analyse in more detail the way in which the size and direction of each source's contribution to overall inequality is determined.

SUMMARY AND CONCLUSIONS

There are numerous possible ways of measuring living standards. There is no single measure that is the correct one, since the very concept of the standard of living that we are trying to capture is a woolly one. Instead, as we have seen in this chapter, there are a number of different measures that we could use. Each one provides information about a different aspect of people's circumstances and gives answers to different sorts of questions we might want to ask. There are a number of approaches to measuring living standards which we use in the pages that follow.

In much of what follows, we settle on one basic *income* measure, based on data from the Family Expenditure Survey. This measure is *current, weekly* income. This provides us with a snapshot view of people's circumstances: their income in the week in which they were interviewed for the FES. This approach is the one followed by the government in its official statistics on low incomes.[26]

The income unit that we use for this is the *household*. Income is added up across all household members, and each individual in the household is assumed to benefit equally from this income. This choice reflects partly the lack of information that we have about the different sorts of sharing arrangements that go on within different households, and partly the fact that it would clearly be misleading to assume that no sharing at all went on between people who live together.

The household's income, for our purposes, is made up from six sources—earnings, self-employment income, private pensions, investment income, social security benefits, and a residual category of other income such as maintenance payments and other allowances and incomes not readily attributable to the other sources. We use an income definition that is *net of direct taxes, National Insurance contributions, and local taxes*. In what follows, we concentrate very much on the measure of income before housing costs (BHC)—that is, total income inclusive of that spent on housing costs—though we do provide some results on the after-housing-costs (AHC) measure when something extra is clearly added to the analysis by doing so.

The income measure we use is one of *equivalent* income—that is, it has been adjusted so that the living standards of households of different size and composition can be compared. This is done using the McClements equivalence scales. All the income figures are presented in terms of the equivalent income of a childless couple.

Our second approach to living standards is introduced in Chapter 4, where we consider the distribution of household expenditure. Because the amount that people spend often reflects their expected resources over the longer term, this provides more of a lifetime view of living standards. The spending measure used for these purposes is closely akin to the income measure used in the previous chapters. It is also based on information from the FES, using current, weekly, household spending. The measure is equivalised in order to compare households of different sizes.

One question that often arises in the analysis of living standards and incomes is 'how do government taxes and benefits affect the distribution of incomes?'. This issue is addressed in Chapter 7. In order to examine the impact of direct taxes in particular, our original definition

---

[26] See DSS (1995).

of income, which was net of such taxes, is no longer sufficient. Alongside this measure, we also look at the amount of direct taxes paid as a proportion of *gross* (that is, pre-tax) income.

Another issue that is often raised in the context of the distribution of income is poverty. Again, income is not the only indicator for assessing poverty. In Chapter 8, we draw from a number of different sources about incomes, spending, and access to consumer durables, and also look at some more qualitative assessments of poverty and hardship.

An important aspect of living standards is *how long* people spend at their level of income. With our original measure of income from the Family Expenditure Survey, we cannot address this directly since the FES is based on a different cross-section of the population in each year. In Chapter 9, we use rather different information to assess the dynamics of income. Using a relatively new data source, the British Household Panel Survey, we are able to see how the incomes of the *same people* change over time.

There are a vast number of possible summary inequality indices of which a small subset are used here. The ones used in this book were set out in this chapter to provide an understanding of what follows, but in no sense to give a full introduction to inequality indices in general and all the other possibilities available. As with the description of the measurement issues, the intention is to provide the tools with which to make the most of the rest of the book.

2 The Current Distribution of Income in the UK

INTRODUCTION

This chapter sets out to provide a detailed description of the current UK income distribution. For this purpose, current means the two years 1992 and 1993, the years for which data were most recently available at the time of writing. The data used are the Family Expenditure Survey, described in the data section of Chapter 1. Here, we concentrate on the distribution of household incomes among individuals, generally treating income as a homogeneous lump and thus not distinguishing between the different income sources. A more disaggregated description showing where the income in each household comes from is reserved for Chapter 5.

The objectives of this chapter can be set out as follows: to convey the shape of the distribution and the numbers of people at each point in the distribution; to give a clear indication of the meaning and extent of the inequality that there is; to show where different sorts of people are in the income distribution; and to show how income is divided unequally not only among the whole population but also among subgroups that might be thought to be rather more homogeneous.

The main income definition used will be an equivalent net household before-housing-costs (BHC) definition where equivalisation is carried out according to the McClements scales described previously. Some reference will also be made to the after-housing-costs definition where appropriate. The advantages and disadvantages of using these income definitions were discussed earlier but for the most part we will write as though they are the only definitions.

Here we look at net (or disposable) income only. The effects of the tax and benefit system and the relationship between original and disposable incomes are discussed in Chapter 7, while the difference between income and expenditure as measures of welfare are explored in Chapter 4. Chapter 8 concentrates on the lowest part of the income distribution, while Chapter 9 demonstrates the effects of taking just a snapshot of income. In this way, our current chapter represents just the first step in our exploration of inequality in the UK.

THE OVERALL DISTRIBUTION

Mean equivalent income (in January 1995 prices) in 1992–93 was £267 per week (£233 on the AHC measure). Remember that this is *equivalent* income, not cash income, so a single person with cash income of £267 would have well above the mean while a couple with children with the same cash income would have less than the average equivalent income. This single number reveals nothing about the distribution of income. It could be equally divided among the population or it could all be held by one person. Add another average to our knowledge about income levels, however, and the amount that we know expands. Median income in the same years was £222, significantly below the mean. This implies that the majority of the population enjoyed an income below the mean level. Those with incomes above the median must on the whole have incomes further from the median than those with incomes below the median.

One could build up quite a detailed picture of the shape of the distribution by now adding to these figures with figures showing the income level at various percentiles of the distribution and the numbers of people with incomes above and below various proportions of the mean. In fact, this is essentially how official statistics are presented in the DSS's Households Below Average Income analysis. Here the compositions of the various income deciles and the numbers of individuals with household income below various proportions of the mean are presented. Before going on to break the income distribution down numerically in this way, we will start by showing graphically what the income distribution actually looks like.

Figure 2.1 shows the number of people with equivalent household income by £5 ranges. The height of each bar shows how many people have a particular level of income. The highest bar is at £120–£125 indicating that nearly 1.25 million people have equivalent household incomes between £120 and £125 per week. The richest 1.6 million people have incomes in excess of £700 per week and do not appear on the graph. In other words, the upper tail of the distribution is much longer than that shown here.

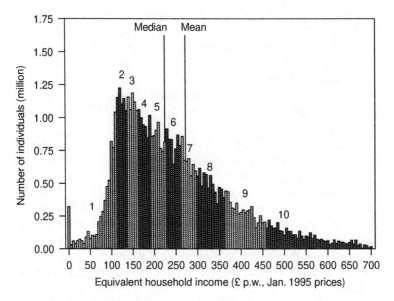

Fig. 2.1. The 1992–93 BHC income distribution

Groups of bars in the graph are alternately shaded light and dark. Each shaded group contains 10 per cent of the population. The light shaded group on the far left of the figure contains the poorest 10 per cent or bottom decile. The next shaded group contains the second decile and so on. Mean income is marked at £267 per week, and median income at £222 which, of course, corresponds with the right-hand side of the fifth decile.

A number of features of the income distribution are at once evident from Fig. 2.1. There are relatively few people with the very lowest incomes of about £70 per week or less. Remember that these are equivalent incomes so that £70 equates to about £43 for a single childless person. Beyond this income, however, the distribution becomes very dense very quickly, with the number of people in each income range rising to a maximum for the whole distribution in the middle of the second decile. The very large numbers of people in the ranges between about £90 and £150 per week reflect levels of social security benefits and the large number of pensioners in receipt of a state pension and just a little other income. At first sight, it appears that even after taking equivalisation into account, this clustering is at slightly too high an income level to reflect benefit rates. This apparent puzzle is clarified, however, by looking at Fig. 2.2, which shows exactly the same information as Fig. 2.1 except for the fact that the AHC measure of income is used. The clustering is both more pronounced and slightly lower down the distribution where one might expect benefit rates to be. The higher level of income in the original figure in fact reflects the inclusion of Housing Benefit in that measure of income.

As one moves right along the horizontal axis in Fig. 2.1, the height of the bars gradually decreases and so the income range that each decile covers grows. While the difference in income between the poorest person and the richest person in the second decile is just £25, the ninth decile covers a range of around £100. Mean income is only reached in the seventh decile—over 60 per cent of the population have incomes below the mean. The much greater compression of incomes below the median by comparison with that above the median, which we predicted from the relative values of the mean and median, is clearly evident.

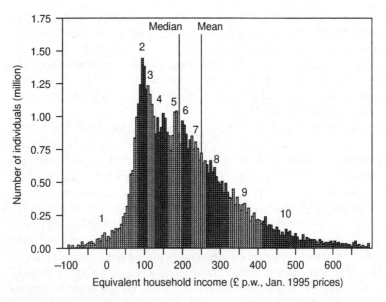

Fig. 2.2. The 1992–93 AHC income distribution

Box 2.1.   Net incomes required by three families by decile
(Figures show maximum net incomes for each decile)

Decile	Single person	Married couple	Married couple with two children (aged 4 and 13)
1	£66	£108	£157
2	£81	£133	£193
3	£97	£159	£231
4	£115	£188	£273
5	£135	£222	£322
6	£157	£257	£373
7	£182	£299	£434
8	£218	£358	£519
9	£278	£456	£661

By the time the tenth or richest decile is reached, incomes are very dispersed. The tenth decile covers an income range from £450 per week right up to the very richest people in the country. The highest-paid executives and entrepreneurs with gross earnings of £1 million per year could have a net weekly equivalent income of about £12 000. It is a point worth stressing that the top decile covers a vast range of incomes and most people in it are far from being among the super-rich. A single person earning £25 000 per year would comfortably make it into the top decile but would have little else in common with those earning hundreds of thousands of pounds a year. The difference between those in the top percentile and those at the lower end of the top decile is considerable—the ninety-ninth percentile income is about 2.3 times as big as the ninetieth percentile income. So while there is a difference in excess of 100 per cent between incomes only nine percentiles apart at the top of the distribution, for someone at the fiftieth percentile a doubling in income would move them up forty percentiles to the ninetieth.

This sort of analysis shows us how many people there are at each point of the income distribution. It provides a clear view of the spread or concentration of incomes. This is essential in the design of a host of government policies, especially those associated with the tax and benefit system. Any tax or benefit related to income will have both its effectiveness and cost (or revenue-raising potential) affected by the way in which incomes are spread. Benefit systems, for example, can create poverty traps whereby very high marginal withdrawal rates of benefits can create serious work disincentives. If these reach up into a very dense part of the income distribution, then they are likely to have a much more unwelcome impact than if they cover only a relatively sparsely populated part of the distribution.[1] The discussion over the appropriate highest rates of tax is often very poorly informed by any clear idea of where the rates should be set. This in turn reflects a very poor understanding of the shape of the income distribution. The idea that a 50 per cent rate of income tax starting at £60000 per year would either affect many people or raise much revenue is fanciful precisely because so few people have incomes that high.

One can also read off a slightly different set of statistics from Figs. 2.1 and 2.2, statistics that show how many people have incomes above and below various proportions of the mean. With mean (BHC) income of £267, half average income is close to £134 and 20 per cent of the population have incomes below this level. Forty per cent of average income is £107 which corresponds very closely to the first decile point. In fact, 9.5 per cent of the population have incomes below 40 per cent of the mean. Such measures can be used as relative measures of poverty or simply as another way of describing total inequality. At the other end of the distribution, 15 per cent of people have incomes over 1.5 times the mean and 6 per cent have income over twice the average—twice the average income is well into the top decile. Even so, nearly 2 per cent or 1.1 million people have incomes over three times the average.

DWARFS AND GIANTS

There is another way of looking at the overall income distribution which shows much the same information as that already described but is much more striking. It involves thinking of a parade of the whole population passing before our eyes in a period of one hour. Each person's height is determined by their income such that the person with the mean income has the mean height (say 5'9"). Some- one with income half the average would have a height half the average, and a person with income twice the average would be

---

[1] Strictly, it is the pre-benefit earnings distribution that matters for this example, but the point remains valid. Dilnot, Kay, and Morris (1984) provide a clear explanation of the issues at stake here.

Box 2.2.   Who would be where in the distribution
(1995 prices)

Bottom decile	Income Support for a single person over 25 was £46.50 per week in 1995–96, for a childless couple £73 per week, and for a couple with two children £122.60. Anyone with just this income would be in the bottom decile. Pensioners with just the basic pension (£58.85 for a single pensioner, £94.10 for a couple) would also be here.
Decile 2	Benefit recipients with some Housing Benefit and pensioners with a small amount of private income would be in this decile.
Decile 3	A single pensioner with just £35 of occupational pension or Housing Benefit would make it into decile 3. They could be joined by a childless couple with one partner earning £10 000 gross per year or a couple with children and one partner earning £15 000.
Decile 4	Single people on very low earnings, pensioners with a little more private income, couples with slightly higher earnings than in decile 3, and some benefit recipients with earnings-related pensions or high levels of Housing Benefit would be in decile 4.
Decile 5	A single person earning just £8500 would make it into decile 5. That is equivalent to about £4 per hour for a 40-hour week and weekly take-home pay of £130. A childless couple with one partner earning £15 000, or a couple with children and each partner earning £10 000, would also be here.
Decile 6	A single pensioner with an occupational pension of £5000 gross would have enough to get into the top half of the income distribution. A couple of low earners on just £8500 each would also make it. But for a couple with children, if only one was earning, gross pay of £23 000 would be required.
Decile 7	For a single person, £12 000 gross would be sufficient to place them in decile 7. The same earnings for a single-earner couple with children would put them in decile 2 or 3.
Decile 8	A single-earner couple with children would need more than £30 000 gross to make decile 8.
Decile 9	Both mother and father would need to earn around £20 000 per year for a couple with children to be in decile 9. A single person on average annual male earnings of around £19 000 per year would very comfortably make it into the upper half of this decile.
Top decile	This decile could include a single person earning £22 000 per year. But for a couple with children, if only one of them was earning, gross pay would have to be in excess of £50 000. A childless couple with each partner earning just £17 000 per year could join them in the richest decile.

twice as tall as the average. Now suppose they pass before us in order of income (and therefore height) with the poorest (shortest) first until the last second of the sixtieth minute when the richest (tallest) person passes.[2]

What does the parade look like? As we have already seen, the majority of the population have incomes below the average; therefore most individuals in our parade will be of below-average height. Indeed, it has been called a 'parade of dwarfs and a few giants'. It will not be until about 36 minutes through that anybody of average height is reached.

The first few seconds will actually see a few upside-down people with negative incomes and therefore negative heights. These will be the self-employed who are making losses from their businesses. And then for the first couple of minutes tiny dwarfs of under a foot or so will be passing. The heights of those passing will initially rise quite quickly, reaching 2'4" after 6 minutes. But then there will be a long parade of dwarfs whose height will increase very slowly, only reaching just over 2'10" (or half average height) by the end of the twelfth minute. Most of those passing at this point will be on social security benefits of some sort.

In the next 18 minutes, taking us up to the half hour, the height of those passing continues to rise gradually, reaching about 4'9" when the half hour is reached. Half-way through and we are still looking down on people nearly a foot shorter than the average. The average height is eventually reached in the thirty-seventh minute, with the height still growing fairly gradually. At this stage, we are seeing mainly working people.

At about the three-quarters-of-an-hour mark, something happens to the parade. The heights of the people passing by start rising much more quickly. It took half an hour for everyone under 4'10" to pass. It then takes another 18 minutes for the height of the parade to reach 7'8". This is the height of people passing with just 12 minutes to go before the hour is up. But over just the next 6 minutes, the height rises to nearly 10', and in the last few minutes, the heights start rising very quickly indeed. By the time we get into the last minute, 15'6" giants are passing by. But it is not until the very last few seconds that the real

[2] This idea was first used by Pen (1971) and is sometimes known as Pen's parade. Other authors to use it include Atkinson (1983), Jenkins and Cowell (1994), and Hills (1995).

giants are striding past. A merchant banker or chief executive of a large company with a gross income of £1 million per year (say) and net earnings, therefore, of about £12000 per week would be towering up in the sky at a mighty 265′ or 88 yards tall, over one-and-a-half times as high as Nelson's Column.

Even above the highest-paid executives and employees will be a few entrepreneurs and aristocrats. The very richest in the country are not salaried employees. They either own their own companies (like Richard Branson) or large parts of the country (the Queen, for example). Unfortunately, none of our data contain information on this particular group of the population. When Pen first wrote of the parade, he estimated that John Paul Getty was the richest man in Britain and attributed to him a height of at least 10 miles.

INCOME SHARES

Yet another way of thinking about the income distribution is to consider the share of income held by each decile. If income were equally distributed, then each decile would hold 10 per cent of the total income. The extent to which income holdings vary from this gives an indication of the degree of income inequality. Indeed, it is on a conception of inequality of this sort that the Gini coefficient (described in the previous chapter)—the most popularly used of all summary inequality statistics—is based. The shares of total income received by each decile are shown in Table 2.1.

The poorest 10 per cent of the population only have around 3 per cent of the total income, less than a third of what they would have if income were equally distributed. Each of the lower six deciles receives less than a tenth of the total income, and they only receive 36 per cent of total income between them. The richest 30 per cent have well over half of all income between them, while the richest 10 per cent alone receive over a quarter of all income.

The shares of income that various groups have tell us about their control over resources. They should also inform us about the way the tax system works. People are often surprised when told that the richest 5 per cent of income taxpayers pay 33 per cent of all income tax, and take this as an indication of the progressivity of the UK tax system. In fact, without a knowledge of the income shares of this richest group, that fact alone tells us nothing about progressivity of the tax structure.

Table 2.1.   Income shares by decile

Decile	Income share (%)[a]	Cumulative share (%)
1	2.95	2.95
2	4.50	7.45
3	5.45	12.95
4	6.50	19.40
5	7.65	27.10
6	8.95	36.05
7	10.40	46.45
8	12.25	58.70
9	15.00	73.70
10	26.30	100.00

[a] Note that this refers to shares of equivalent income. In a sense, it is a measure of the share of 'worth' of money, not just money income itself.

Note: The decile shares may not sum exactly to the cumulative share due to rounding.

Given what we know about the proportion of total income received by the richest group, it is not so surprising that they pay such a large proportion of total taxes. It is to a large extent because they receive such a large proportion of total income.

AGE

In Chapter 1, we constructed a possible world in which one might observe a great deal of inequality just because people were captured in the snapshot picture of the income distribution at different points of their life cycle. So the natural first question to ask is 'to what extent is the degree of observed inequality simply a result of inequality between people of different ages?'.

If one divides the population into 10-year age-groups, divided according to the age of the head of the family unit, one actually sees very little variation in the level of average income between those in their 20s and those in their 50s. The mean income for those in their 20s is around £260 per week, rising to £300 for those in their 40s, and only falling back £5 for those in their 50s. There is, though, a fall between this age-group and the over-60s, and a further big fall for the

Table 2.2.   Distribution of age-groups by quintile

Age-group	Quintile 1	Quintile 2	Quintile 3	Quintile 4	Quintile 5
<21	23	19	24	21	13
21–30	20	19	18	21	21
31–40	22	17	20	20	21
41–50	14	15	21	23	26
51–60	17	14	20	24	25
61–70	19	28	22	17	14
>70	28	36	17	11	8

oldest groups such that the mean income among those over the age of 70 is just £200—about two-thirds that of the richest age-group.[3]

There are two reasons for differences between incomes among different age-groups: the first is the life-cycle effect already discussed and the second is a cohort effect. The latter refers to differences between groups of people born at different times—differences that arise because of their generation, not their age. This latter effect can arise from changes in the economy or the sorts of skills that different generations possess. It is discussed in more detail in Chapter 9 dealing with dynamic issues. We raise this issue here just to make clear that even to the extent that inequality does arise from differences between different age-groups, some of it can arise from genuine lifetime differences in incomes between different age-groups as well as life-cycle effects which obscure a real similarity between age-groups. Indeed, decomposable inequality indices, such as the mean log deviation, suggest that almost all observed inequality can be accounted for by differences *within* age-groups rather than between them.

To complete this brief description of the age-related aspects of inequality, we show in Table 2.2 how the different age-groups are distributed across the population. We divide the population into fifths (quintiles) from the poorest to the richest and calculate what proportion of each age-group falls within each quintile. A completely even distribution by age-group would result in 20 per cent of each age-group being found in each quintile group. Among those in their 20s

---

[3] For each group, the median is approximately £40–£50 less than the mean—the relativities do not change if this measure of the average is used instead.

and 30s, the evenness of the spread among the quintiles is remark-able—near enough a fifth of them are found in each quintile. The greater average prosperity of people in their 40s and 50s is clear from the lower proportions of them in the bottom quintiles and the fact that around a quarter of them are to be found in the richest group. But of the oldest group, nearly two-thirds are to be found in the poorest 40 per cent of the population, with especially great over-representation in the second quintile. This oldest group is only half as likely as the average to be found in the richest 40 per cent of the population.

FAMILY TYPES

The breakdown by age-group is only one of numerous possible ways of decomposing the population to reveal its constituent parts and their contribution to the overall picture of inequality. In this section, the population is broken down by family type, and the distribution of different types of family within the overall income distribution is examined. The distribution of income within each family type is also considered. In the sections that follow, similar analyses dividing the population according to economic activity, to region, and to housing tenure status will be presented.

Before breaking the population down by family type, it is worth quite simply looking at the proportion of men, women, and children at various points in the income distribution. We still do this using a household measure of income, therefore assuming equal sharing *within* households. So even if a man in a household is earning a large salary and a woman in the same household is not earning, they are both assigned the same income.[4]

Figure 2.3 shows the composition of each decile broken down between men, women, and children. Three patterns are immediately obvious. First, children are over-represented near the bottom of the distribution and under-represented near the top. They account for just over 30 per cent of individuals in the poorest decile and only 14 per cent in the richest decile, as against around 23 per cent in the popula-tion as a whole. It is not, perhaps, surprising to find children on average in families towards the lower end of the income distribution.

[4] Webb (1993) and Sandra Hutton (1995) look in detail at the individual distribution of incomes of different sorts between men and women.

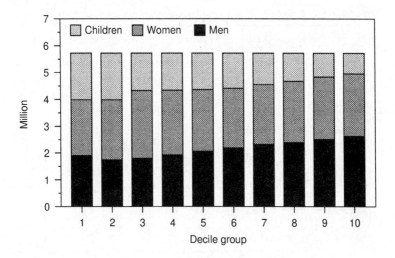

Fig. 2.3. Numbers of men, women, and children in each decile

This is partly because such families are less likely to have both adults (or the only adult) in full-time work than is the case among the population as a whole, and partly because the presence of children means that more raw income is needed to achieve the same living standard and therefore the same level of equivalent income. If a child is born to a family, then, other things being equal, the family's equivalent income will fall. Naturally, this means the results are likely to be sensitive to the particular equivalence scale used to normalise incomes. In fact, the McClements scale used throughout this study gives weights to children that are, if anything, lower than the average in other scales. Using a scale with greater weights would increase the concentration of children near the bottom of the distribution.

The second obvious trend is the increasing proportion of men in each decile as we move up the distribution. Men make up 45 per cent of the top decile and only 33 per cent of the bottom decile. Women, by contrast, are fairly evenly spread through the distribution. The result is that women outnumber men by 1.1 to 1 in the poorest decile while men outnumber women by a similar proportion in the richest group. The biggest gaps are in fact in the second and third deciles, where women outnumber men by 1.3 to 1 and 1.4 to 1 respectively.

Table 2.3. Distribution of different family types by quintile

Family type	Percentage of population	Quintile 1	Quintile 2	Quintile 3	Quintile 4	Quintile 5
Couple pensioner	9	25	29	21	14	11
Single pensioner	8	25	40	16	10	8
Couple with children	38	20	18	24	21	17
Couple, no children	22	10	11	17	26	36
Single with children	7	42	33	13	8	4
Single, no children	16	18	16	20	23	23

This reflects a number of factors including the high proportion of women among the very old (resulting from their greater longevity) and the low incomes of single-parent families, the vast majority of which are headed by women. The relatively high proportion of men in the poorest decile by comparison with the next two deciles is a reflection of the very low incomes of the unemployed and some of the self-employed. These are more likely to be men than are the pensioners, for example, who make up a large part of the second decile.

The remainder of this section looks at particular family types rather than at individuals by sex. Table 2.3 breaks the population down by family type according to whether or not over state pension age, whether married[5] or single, and whether or not there are children in the family. Like Table 2.2, it shows how each family type is distributed among the income quintiles. Also shown is the overall make-up of the population. The fact that 30 per cent of a family type are found in any quintile does not mean that it forms the major component of that quintile because there is not an even distribution of family types. There are many more couples with children in the population than there are single people with children. And so, even where single parents are over-represented and couples with children under-

---

[5] Married includes cohabiting for these purposes.

represented, the latter always outweigh the former in terms of raw numbers. The figures we present, therefore, show how each family group is distributed through the income distribution rather than how each income group is composed by family type. (Of course, there is enough information in the table to construct that easily enough.)

The patterns are easy enough to pick out. There are two groups over-represented at the bottom of the income distribution—pensioners (both married and single) and single-parent families, 42 per cent of whom are to be found in the bottom quintile. Although we divide the population only into quintiles for the purposes of this analysis, largely to help exposition, it is worth considering separately the two deciles that make up the bottom quintile. For the bottom decile and the second decile are actually quite different. Pensioners are not over-represented in the poorest decile, where fewer than 8 per cent of pensioner couples are found, but are massively over-represented in the second decile. By contrast, couples with children are over-represented in the very poorest group but not in the second decile. These patterns are the result of a number of circumstances which will become clearer below where a similar analysis is carried out with the population divided by economic status. Essentially, the bottom decile contains many of the unemployed and self-employed who are poor enough to push many of the pensioners just out of the bottom decile and into the second decile. In none of the other quintiles is there this significant difference between the two constituent deciles.

These bottom two deciles are also the only place where it makes a big difference whether we use income before or after housing costs. The BHC figures are the ones shown in Table 2.3. On the AHC basis, pensioners are barely over-represented in the bottom quintile and significantly under-represented in the poorest decile. The relative position of couples with and without children worsens, couples with children being significantly over-represented in the poorest decile on this measure of income. This difference at the bottom is largely attributable to the role of Housing Benefit, which counts as an income in the BHC measure but effectively does not in the AHC measure. Pensioners tend to have lower housing costs, and more low-income pensioners are outright owner-occupiers, than is the case among non-pensioners. Hence the BHC income measure probably exaggerates the income levels of non-pensioners relative to those of pensioners. For those at the bottom of the income distribution, the AHC measure is probably a better one.

For the rest of the distribution, we need worry neither about differences between BHC and AHC measures nor about deciles rather than quintiles. Little extra information would be gleaned from extending the analysis to cover these groups separately.

The difference between the top quintile and the bottom one is striking, but not surprising. Lone parents are barely represented at all at the top of the distribution, only 4 per cent making it into the top quintile. Pensioners are also under-represented, with 11 per cent of couple and 8 per cent of single pensioners in this quintile. Couples with children are found predominantly in the middle parts of the distribution. The richest group is clearly the childless couples, over a third of whom are in the top quintile. This reflects both straightforwardly their lack of children, meaning that their incomes are not equivalised downwards, and the greater likelihood of both partners being in work.

The distribution of families across the overall income distribution gives a good impression of which family types are typically better off and which worse off. We just confirm that pattern if we look at the median incomes of the various groups. The differences are significantly greater than those by age-group, and partly explain why the most inequality is within particular age-groups and not as a result of differences between them.

Both the richest and poorest groups—childless couples and single parents—are below pension age. With a median equivalent income of around £300 per week, the former group are on average more than twice as well off as the latter group, who have a median income of just £140. Pensioners are significantly better off than single parents but still have average incomes considerably below those of the other working age-groups. The median equivalent income for pensioner couples of £180 and for single pensioners of £160 compare with £220 for couples with children and £240 for childless single people.

Within-Family-Group Inequality

The figures so far presented reveal much about where different family groups are in the overall income distribution and how average incomes differ between them, but relatively little about inequality within each group. As we confirm in Chapter 6 in which we try to explain the observed patterns of inequality and how they have changed, most of inequality is caused by within-family-group differences rather than by differences between them.[6]

At the bottom end of the distribution, the poorest members of each group have remarkably similar incomes. With the exception of the childless couples, the family 10 per cent of the way up each family type distribution has an income within about £10 of that enjoyed by the family in the equivalent position in each of the other family type distributions. So, for example, the tenth percentile single pensioner has £107, the equivalent single parent has £99, and the couple with children has £103. This similarity at the bottom must reflect social security benefit levels which, roughly speaking, will leave the poorest groups on a par with each other. As one moves further up the distribution, the families move apart. At the twenty-fifth percentile, the single parents have £18 per week less than the pensioners, who in turn have £20 per week less than the childless single people. The twenty-fifth percentile childless couples with £206 per week are £50 better off than the childless single people.

The differences are most marked at the top of the piles. The ninetieth percentile childless couple, with £570 per week, is £100 better off than the equivalent childless single person, who has £100 more than married pensioners in a similar position, who are in turn £100 better off than the richest single parents.

To look at the within-group distributions more formally, the next stage is to compute the Gini coefficient for inequality within each of the family groups. This is displayed in the 'Hasse' diagram in Fig. 2.4. This figure should be interpreted in the following way. The group at the top has the lowest Gini. Where a line runs down from one group to another, there is unambiguous Lorenz dominance. In other words, the higher group displays unambiguously less inequality. Where there are no lines running between groups, it is not possible to make unambiguous statements regarding relative degrees of inequality, though of course comparisons can still be made between Ginis.

From the diagram, it is clear that lone-parent families have the lowest Gini by some distance. They are unambiguously more equal as a group than all the other groups bar pensioner couples and the single childless. The two pensioner groups have the next lowest Ginis, but cannot be ranked unambiguously. Both are unambiguously more equal than couples with children, and married pensioners are unambiguously

---

[6] See, for example, Jenkins (1995) and Goodman, Johnson, and Webb (1994).

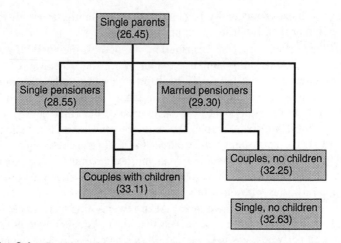

Fig. 2.4. Ranking of inequality among family types by Gini coefficients

Note: Numbers in parentheses are Gini coefficients. Where there is a vertical line between two groups, the higher group is unequivocally more equal than the one beneath. Where there is no line connecting, no unequivocal Lorenz ordering can be made.

more equal than childless couples. Overall, there are only a limited number of complete comparisons that can be made, there being some crossing of Lorenz curves in nine out of the possible fifteen pairwise comparisons.

In fact, there is no unambiguous comparison at all that can be made between the single childless and any other group. This is why they are sitting outside the main part of the figure unconnected with any of the other groups. This is a peculiar situation, which arises despite a Gini more than six points higher than that for lone parents. Their Lorenz curve is outside that of every other group until past half-way up the income distribution. At various percentiles, it then moves inside—at the fifty-ninth when compared with married with children, the seventy-sixth against childless couples, and the eightieth, eighty-fifth, and ninetieth when compared with single pensioners, married pensioners, and lone parents respectively. This pattern indicates that while poorer single people have a small share of total income, this reflects a high share among the middle groups but a relatively low share for the richest single people.

Even lone parents turn out not to be unambiguously more equal than all the other groups on this basis. Their Lorenz curve moves outside that of couple pensioners at the ninety-second percentile and that of the single childless at the ninetieth percentile. This must reflect the fact that there are a very few rich lone parents such that the very top percentiles have a somewhat higher proportion of the group's total income than do the very richest single childless and married pensioners. But in the bulk of the distribution, their incomes are very equally spread and, until the very top is reached, the cumulative income shares are high.

Ethnic Minorities

There is a final demographic breakdown which we are unable to study in detail given the data available to us, and that is a breakdown by ethnic group. The FES data simply contain no information on this. However, a limited amount of information is available from other surveys, in particular the General Household Survey (GHS). In general, we do not use this survey for looking at the distribution of income because the income information in the GHS is not as comprehensive as that in the FES. Nevertheless, the GHS can provide some information about the position of ethnic minorities. In particular, it reveals that over a third of those classified as non-white are to be found in households with gross equivalent incomes in the lowest fifth. Only just over 10 per cent are in the richest fifth. If one looks only at those aged under 65 (non-whites are relatively unlikely to be of pensionable age where incomes on the whole are lower), one finds that non-whites are twice as likely as whites to appear in the bottom quintile and half as likely as whites to appear in the richest quintile (Hills, 1995).

The ethnic minority population is itself diverse, and different ethnic groups within it appear to have very diverse experiences. GHS data suggest that those of Indian origin are actually significantly under-represented in the poorest two quintiles and over-represented in the third and fourth quintiles. The West Indian population appears to be rather poorer than this, with substantial over-representation at the bottom, though with still nearly 20 per cent in the third and fourth quintiles, while Pakistanis and Bangladeshis appear to be almost uniformly poor, nearly 60 per cent being found in the lowest income quintile with over 25 per cent in the second quintile.[7]

[7] Source: Hills, 1995.

ECONOMIC STATUS

As has already been stressed a number of times, those near the bottom of the income distribution are largely dependent on social security benefits for their incomes. They have no significant income from employment. And as became clear during the discussion of the position of various family types within the income distribution, it is very much employment position which is instrumental in determining where a family falls within the overall distribution. Here, we look in some detail at the role of economic status in determining where people fall within the overall distribution.

We start by looking at the proportion of adults in each decile who are in work. This shows very clearly the strong relationship between income and employment status. Figure 2.5 shows that in the bottom two deciles, just under a quarter of adults are in work. This climbs to a third in the third decile and a half in the fourth decile, before reaching just over 80 per cent in the top two deciles.

There are a number of possible ways of splitting the population down by economic status. Here, as in DSS (1995), the population is

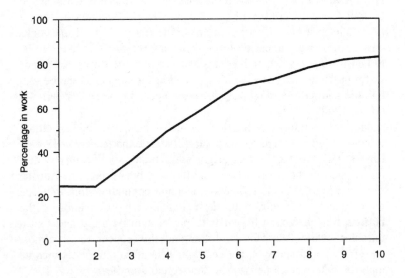

Fig. 2.5. Percentage of adults in each decile in work

split into eight economic status subcategories. The proportion of each in each quintile of the population is shown in Table 2.4 in the same way as different family groups are shown in Table 2.3.

As one would expect, the unemployed and 'other' categories are very much over-represented at the bottom of the distribution. Over two-thirds of the unemployed and approaching half the 'other' group are in the poorest quintile. Looked at the other way round, between them they account for half of the poorest decile and with the over-60s they account for 69 per cent of the poorest quintile. By contrast, all families in which there is a full-time worker are under-represented at the bottom. Only 1.5 per cent of all families containing only full-time workers are in the bottom quintile. Even among one-earner couples, 11 per cent are in the poorest group. Between them, families containing at least one full-time worker account for half of the population but only 12 per cent of the poorest quintile (excluding the self-employed). The message is clear. Having a family member in work, and in particular in full-time work, is the best way of keeping out of the bottom fifth of the income distribution. Having all adult family members in full-time work is an almost foolproof way of staying out of the poorest 20 per cent. The contrast with the position of the unemployed is striking.

Naturally, the top end of the income distribution simply displays the other side of this particular coin. There are virtually no unemployed right at the top of the income distribution. Almost 90 per cent of those families in the top quintile contain at least one full-time worker (including self-employed). Three-quarters of families containing only full-timers are to be found in the top two quintiles. The increasing representation of workers and falling representation of non-workers as one moves up the distribution is in fact shown quite clearly in Table 2.4. The proportion of the unemployed in each quintile falls quickly, until fewer than 2 per cent are to be found in the top quintile. The over-60s also fall away, though not so dramatically. Conversely, it is only when the fourth quintile is reached that families consisting wholly of full-time employees become over-represented.

One other group is worthy of note in its own right as it displays characteristics not shared by the others examined—that is the self-employed. Although the table is not adequately disaggregated to show it, 16 per cent of them are in the poorest decile, though the fact that relatively few are in the second decile results in them not being

Table 2.4. Distribution of different economic status groups by quintile

Economic status	Percentage of population	Quintile 1	Quintile 2	Quintile 3	Quintile 4	Quintile 5
Self-employed	10	22	16	18	20	24
All full-time	22	2	6	17	32	43
One full-time, one part-time	13	2	15	33	28	22
One full-time, one not earning	14	11	22	26	23	18
All part-time	7	27	28	21	13	12
Over 60	18	27	34	19	12	8
Unemployed	7	69	17	8	5	2
Other	9	46	30	13	7	5

greatly over-represented in the poorest quintile. Fewer than 20 per cent of the self-employed are to be found in each of the succeeding quintiles until the richest group is reached, where a quarter of the self-employed are to be found. Again, there is a divergence between the two top deciles. Slightly fewer than 9 per cent of the self-employed are in the ninth decile but nearly 15 per cent are to be found in the richest decile. There seems to be a considerable divergence in the experiences of the self-employed, with a large group of them having, or claiming to have, very low incomes indeed[8] and another significant group appearing to be making very substantial profits from their businesses.

This divergence of experience among the self-employed is immediately evident from looking at the Hasse diagram in Fig. 2.6. The self-employed have a very high Gini and are as a group more unequal than any other. With equal lack of equivocation, we can say that that small group of families containing only part-time workers is the second most unequal. Working from the other end, the group with the lowest Gini is families containing one full-timer and one part-timer, but the Lorenz curve of this group does cross those of the unemployed and of full-time workers, which also have crossing Lorenz curves. The difference between the all full-timers group and the one full-time, one part-time group is interesting. The Lorenz curve of the former crosses that of the latter once, from above, at the sixtieth percentile. This means that the full-time group have more equal incomes in the lower part of the distribution, but more unequal ones nearer the top.

Single-earner couples appear to have significantly more unequal incomes than families where all members are at work, reflecting the fact that significant numbers of one-earner couples are to be found throughout the income distribution, as was seen in Table 2.4. The lack of direct comparability with the all full-time group reflects a crossing of Lorenz curves at just the fifth percentile.

Given that we know from much previous work[9] that the wage distribution has widened greatly since the mid-1970s, it is possibly

---

[8] There is some reason to be rather suspicious of the zero, and indeed negative, incomes reported by some of the self-employed at the bottom of the income distribution. On average, they seem to have expenditures above the average for the population as a whole. Their nil or negative reported profits do not seem to be a good indication of their living standards. This issue is touched upon in Chapters 1, 4, and 8. See also Davies (1995).

[9] For example, Machin (1996) and Gosling, Machin, and Meghir (1994).

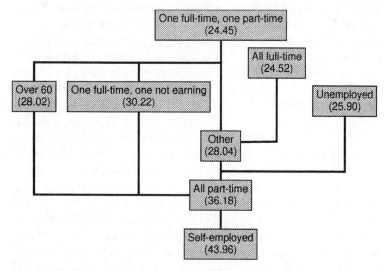

Fig. 2.6. Ranking of inequality among economic types by Gini coefficients

surprising to find full-time workers with the most equal incomes. It is, however, simply a reflection of the fact that families in which there are only full-time workers are very rarely found very near the bottom of the income distribution. A single person with gross earnings of just £10000 per year would have a net equivalent weekly income of nearly £250 per week, just below the mean for the whole population and well above the median. As soon as one introduces non-employed adults into a family, the likelihood of their equivalent income being substantially lower is greatly increased. This explains the greater inequality found among families where not all are in employment.

A final point regarding the inequality indices relates to the comparison between the unemployed and 'other' groups. The latter have a significantly higher Gini, but again the Lorenz curves cross. In this case, that for the unemployed crosses that for the 'other' group from below at the twentieth percentile. This must reflect a number of the unemployed being particularly poor.

Again, we illustrate the inequality within each group by looking at the incomes at particular percentile points of the income distribution for each group (Table 2.5). As with the family types, each economic status group displays major differences between richest and poorest. The very low incomes of the poorest self-employed and very high

Table 2.5. Percentile points for each economic status group

Percentile	Self-employed	All full-time	One full-time, one part-time	One full-time, one not earning	All part-time	Over 60	Unemployed	Other
5	24	167	153	118	83	96	48	72
25	147	252	212	178	140	131	96	116
50	236	327	263	237	190	165	122	146
75	362	427	349	322	272	232	168	203
95	888	660	567	558	509	415	301	392

incomes of the richest are clear, as are the relatively uniformly low incomes of the unemployed. The ninety-fifth to fifth percentile ratio for the one full-time, one not earning group is higher than that for the all full-time group—4.7 as against 4.0. The former group appears to contain both couples with one very high earner and couples with quite low earnings.

REGION

The most natural way of breaking the population down for the purposes of understanding the income distribution is by family type and economic status as we have done. There are other disaggregations that might also give us useful information. Here we describe briefly two further breakdowns that are of interest. In this section, we look at a regional description of the income distribution; in the next section, we break the sample down by tenure type. In neither case do we go into the sort of detail presented for the economic status and family type breakdowns shown earlier.

All sorts of characteristics vary by region, including health status, unemployment levels, and living standards. More pertinently from our point of view, smaller localities within larger regions can be very different from one another. In January 1995, the claimant unemployment rate was 6.3 per cent of the work-force in North Yorkshire and 12 per cent in South Yorkshire, 5.4 per cent in Powys and 11.1 per cent in Gwynedd, 5.8 per cent in West Sussex and 10.5 per cent in East Sussex.[10]

Our data do not allow us to get down to such local levels. Other authors have looked at local authority level and even at unemployment rates at ward level. Green (1994) used census data to divide the population into 459 local authority districts and about 10 000 wards or neighbourhoods. This did not allow income levels to be determined directly but allowed such indicators of income as social class, inactivity rates, car ownership, and education levels to be examined. The localisation of deprivation and affluence is very clear from her work. On all three of her measures of deprivation, Glasgow city is in the top ten, but on two out of three measures of affluence, Bearsden and Milngavie, on the outskirts of Glasgow, are in the top ten. London

---

[10] Source: *Employment Gazette*, March 1995.

boroughs such as Hackney and Tower Hamlets were among the most deprived, but Richmond upon Thames and Kensington and Chelsea scored high on affluence. Other researchers have found significant polarisation between very local parts of towns.[11]

We are constrained by our data to look at much bigger regions. In fact, although we could split the UK into twelve regions, we have decided in the interest of data accuracy to concentrate on four English regions—North, Midlands, South, and London[12]—plus Wales and Scotland. From our data, we see the North accounting for 26 per cent of the population, the Midlands for 16 per cent, the South for 31 per cent, and London for 10 per cent, while Scotland and Wales contain 9 and 5 per cent respectively. We do not provide figures for Northern Ireland because of the small number of observations in the data for inhabitants of that province.[13] It is certainly worth bearing in mind that these large regions contain many subregions and areas that are quite different. Indeed, at this level it is clear that most inequality occurs within the regions, *not* between them.

Table 2.6 shows mean and median incomes for each region. As one would expect, London and the South have the highest average incomes, while Wales has the lowest, with a more than 20 per cent difference between the medians in the poorest and the richest regions. To some extent, these figures overestimate differences in living standards between the regions as some commodities, especially housing, are more expensive in the more affluent regions. Borooah *et al.* (1996) have in fact investigated such differences and, according to their estimates, incomes in Greater London 25 per cent above the UK average actually translate to price-adjusted living standards less than 20 per cent above the average. Conversely, the incomes of the Welsh would be translated from 12 per cent below the average to less than 9 per cent below.

Although not shown here, on the AHC measure of income the differences in the averages are smaller but the pattern is the same. London and the South remain substantially richer than the rest of the country.

The table also shows the Gini coefficient for each region. For most it is clustered around 31 per cent, but it reaches 34 per cent in the

---

[11] Noble *et al.* (1994) looked at patterns of income within Oxford and Oldham.

[12] The North equates to FES standard regions North, North West, and Yorkshire; the Midlands comprises East and West Midlands; and the South is made up of East Anglia, the South East, and the South West.   [13] Though see Borooah *et al.* (1996).

Table 2.6. Regional incomes and inequalities

Region	Mean	Median	Gini
North	241	206	31.50
Midlands	241	207	30.98
South	294	246	34.33
London	333	255	40.12
Scotland	251	219	30.74
Wales	234	201	31.60

South and 40 per cent in London. This makes London really quite different from the rest of the country, reflecting the combination of a relatively large number of unwaged individuals and the fact that many of the earners have very high earnings. The difference between London and the rest of the country is confirmed by the fact that the Lorenz curve for London lies everywhere outside that for every other region. Furthermore, that for the South, while lying inside London's, is also outside that of all the other regions. London and the South are richer than the rest of the country but they are unequivocally less equal.

The most equal region according to the Gini appears to be Scotland, but it is not very different from most of the other regions.

Figure 2.7 shows the percentage of the inhabitants of each region in each income quintile. The Welsh are the most over-represented at the bottom and under-represented at the top. Nearly a quarter of them are in the poorest quintile and only 14 per cent are to be found in the richest group.[14] Southerners are the least likely to be found in the bottom quintile, but Londoners are by a long distance the most likely to be in the highest quintile—30 per cent of them are there. London seems to have something of a hollow middle—only 16 per cent of Londoners are in the middle quintile and they are also under-represented in quintiles 2 and 4. The Scottish are most over-represented in the middle and the most evenly spread across the distribution.

The existence of large numbers of individuals near the bottom of the distribution is clearly correlated with levels of unemployment and regions of relative industrial decline. The particular situation of Wales

[14] In fact, the Northern Irish are even more likely to be in the poorest decile. From our sample, it appears that a third of them are in the poorest quintile with 13 per cent in the richest, but again the small sample size means that there is some uncertainty about that result.

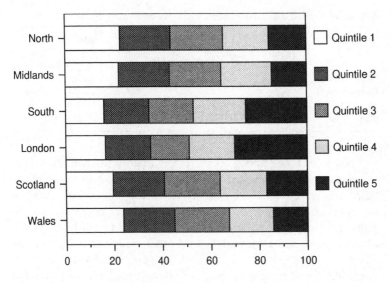

Fig. 2.7.   Percentage of each region in each quintile

has been highlighted in other contexts—there seem, for example, to be a disproportionately large number of Invalidity Benefit recipients there.[15]

HOUSING TENURE

So far, the dominant characteristic of individuals that has affected their place in the income distribution has been their employment status. Another characteristic that is closely correlated with people's place in the income distribution is their housing tenure, though of course this relationship is not causal in the same way as that between employment status and income position is. Indeed, the direction of causality is likely to be the reverse, if anything. One can see how strong is the relationship between income and certain types of tenure from Table 2.7.

[15] See Disney and Webb (1991).

Table 2.7. BHC and AHC means, medians, and Ginis for each tenure

Tenure	BHC mean	BHC median	AHC mean	AHC median	BHC Gini	AHC Gini
Social renters	164	146	134	113	22.16	26.95
Private renters	235	187	187	148	33.92	41.88
Mortgagers	323	276	277	237	31.61	34.65
Outright owners	259	215	255	210	34.04	34.53

Average incomes among social renters (that is, council and Housing Association tenants) are very much lower than those among the other groups, especially on the AHC measure in which Housing Benefit is effectively netted off income. Private renters form a middle group between the social renters and owner-occupiers, but they are also noticeably more unequal than the other groups, especially on the AHC measure of income. Some in the private rented sector are very much like social renters, especially some of those who still have old tenancies covered by fair rent agreements and those others housed in the private rented sector but whose rents are effectively being paid by Housing Benefit. Others in this sector are much more like owner-occupiers, though younger. The greater inequality on the AHC measure will reflect the fact that Housing Benefit was bringing the incomes of the poorer renters up nearer those of the richer ones.

Those with mortgages have the highest incomes on both measures, and are significantly further ahead of outright owners on the BHC measure. Mortgagers have considerably higher levels of economic activity than any of the other groups, including outright owners who tend to be more elderly.

In Table 2.8, we break with the convention of the rest of this chapter and present the main results of the distributional analysis by tenure on the AHC measure of income. For this particular analysis,

Table 2.8. Percentage of each tenure type in each AHC quintile

Tenure	Quintile 1	Quintile 2	Quintile 3	Quintile 4	Quintile 5
Social renters	43	32	15	8	2
Private renters	32	22	16	15	14
Mortgagers	11	14	22	26	28
Outright owners	12	20	23	21	23

this seems to be the appropriate measure, preventing Housing Benefit from making renters appear better off than outright owners with the same other income. On the BHC measure, outright owners look significantly worse off relatively and the big apparent gainers are private renters, fewer than a quarter of whom appear in the bottom quintile on that measure as against a third on the AHC measure.

The over-representation of social renters in the bottom income quintiles is clear. Over 40 per cent of them appear in the bottom quintile and a further third in the next quintile. Only 10 per cent of social renters are to be found in the top 40 per cent of the population. Only a tiny proportion make it into the richest quintile.

Most over-represented at the top of the distribution are owners with a mortgage. This group tends to be comprised mainly of working families. Outright owners are rather less likely than this group to be at the top of the income distribution because a large proportion of them are relatively old.

As we will see in the next chapter, this concentration of social tenants at the bottom of the income distribution is a relatively recent phenomenon. Where social tenants live in large concentrations, this is likely to result in local geographical concentrations of the poor, as documented by Noble et al. (1994). This in turn, combined with the poor quality of much rented housing, will lead to a concentration of deprivation and all the increased needs and problems that are implied by it. Among social tenants under pensionable age, one-fifth are single parents; fewer than a half are in work.

CONCLUSIONS

Inequality in the UK is multifaceted. The overall degree of inequality was best illustrated in Figs. 2.1 and 2.2, which showed the shape of the income distribution. The rest of the chapter has concentrated on looking at various parts of the income distribution broken down in different ways, at how the different parts of the distribution are constituted, and how unequal the groups are themselves.

A number of major conclusions can be drawn from this description of the distribution.

1. There is a considerable bunching of people on relatively low levels of income at and around benefit levels. At income levels higher

than this, there is a gradual thinning of the distribution—each decile covers a wider range of incomes than the one before it.

2. Income is unequally shared. The poorest decile have just 3 per cent of total equivalent income, the richest decile more than a quarter.

3. Women and children are more likely to be found towards the bottom of the distribution than men. Pensioners are found near the bottom. Single parents tend to be the poorest family group, childless couples the richest.

4. But incomes within each family group are very unequally distributed.

5. Economic status is a major determinant of income. Not surprisingly, non-workers tend to be a good deal poorer than workers. Again, each employment group is unequal, the self-employed particularly so.

6. London and the South are both richer and more unequal than other parts of the country. Wales is the poorest part of Great Britain.

7. Social renters are very concentrated near the bottom of the income distribution.

3 Changes in the Income Distribution

The distribution of income changes over time. The degree of inequality changes, the composition of poorer and richer groups changes, and the position of various groups in the income distribution changes. In the main part of this chapter, the emphasis is on describing in some detail the way in which the income distribution changed between 1961 and 1993 using the same type of data and methodology used in the previous chapter. This is the longest period for which data are available on a consistent basis. In the first part of the chapter, however, some attempt is made to draw on other data to put this period into context.

The main chapters (5, 6, and 7) attempting to *explain* what has happened to income inequality come later. So while some commentary on why the changes that are observed might have occurred is included here, the main focus is on description rather than explanation.

SOCIAL AND ECONOMIC CHANGES

To be clear about how the income distribution has changed requires consistent data to be available. We have such consistent data on an annual basis starting in 1961. There is no directly comparable information from before that date, though the existing evidence led the Royal Commission on the Distribution of Income and Wealth to conclude in 1979 that

The overall impression from the figures is of a reduction in inequality but, if the decline in the share of the top 1 per cent is ignored, the shape of the distribution is not greatly different in 1976–77 from what it was in 1949 . . .

The income distribution shows a remarkable stability from year to year. (Royal Commission on the Distribution of Income and Wealth, 1979, p. 17.)[1]

No royal commission reporting today could have been responsible for that last sentence.

Before beginning the description of the changing distribution over the past three decades, it is useful to put what has happened to incomes into the context of what else has happened in the UK. At the start of the 1960s, the UK still had what can reasonably be called full employment with fewer than half a million unemployment claimants. Harold Macmillan was still Prime Minister, having recently told the nation that 'you have never had it so good'. The unemployment level rose very slowly over the 1960s, jumped to a million following the first oil shock in the mid-1970s when Harold Wilson was into his fourth term as Prime Minister, hit 2 million at the beginning of the 1980s following the second oil shock and Margaret Thatcher's election in 1979, and had risen to 3 million by the middle of that decade. A sharp fall during the boom of the late 1980s took unemployment down to about 1.5 million, but an equally sharp rise during the recession of the 1990s saw it rise towards 2.9 million in 1993, the last year for which the analysis is carried out. This pattern of unemployment changes is graphed in Fig. 3.1. Through this chapter, it will be clear where unemployment has affected overall inequality, and Chapter 6 will look more carefully at the role it, among other things, has played in changing the level of inequality.

Unemployment is only one of many things that have changed and affected the distribution of income. There have been three recessions (years in which real GDP has fallen) over the period under consideration: in 1974–75, in 1980–81, and in 1991–92. There have also been boom years of exceptional GDP growth, 1973 being the most exceptional. These troughs and booms in the overall economic performance of the country have had their own effects on the income distribution, effects that have not necessarily been the same in each boom or in each recession. And the effects have not always been what might have been predicted; booms do help those on the lowest incomes by reducing unemployment, but they can also be associated with very high earnings increases for others, especially where they precipitate skill

---

[1] Quoted in Atkinson (1983).

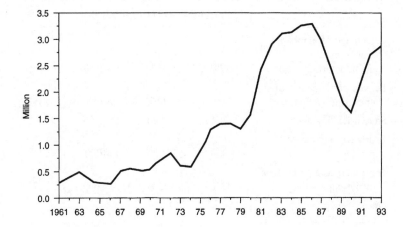

Fig. 3.1. Changing unemployment levels

Note: Claimant unemployment excluding school-leavers; current basis back to 1971.
Source: *Economic Trends.*

shortages as happened in the late 1980s when swiftly falling unemployment was correlated with swiftly rising inequality.

Table 3.1 shows a series of other economic and social changes during the period under discussion. Even among those who have been employed, there have been significant changes in composition. At the start of the 1960s, well over a third of workers were involved in manufacturing. Industrial restructuring has been swift, especially since 1971, since when the proportion of the work-force involved in manufacturing industry has fallen from a third to less than a fifth. At the same time, there has been significant 'feminisation' of the work-force. In 1961, a third of paid workers were women. By 1991, this had risen to 45 per cent, and in the near future it is expected to top 50 per cent. Increased participation by women will clearly have a substantial effect on inequality. Harkness, Machin, and Waldfogel (1996) argue that the effect has been to mitigate rises in inequality since the late 1970s because most of those women moving into employment have been married to men on relatively low wages.

The other changes shown are largely social and demographic ones that have had an effect on income levels and inequality. The number

Table 3.1.   'Then and now'

	1961	1971	1981	1991
Percentage of employed work-force in manufacturing[a]	37%	33%	26%	18%
Female participation in employed work-force[a]	34%	36%	40%	45%
School-leaving age	15	15	16	16
Estimated number of lone parents	468000	570000	900000	1300000
Pensioners as a percentage of the population	13%	15%	16%	17%
Owner-occupier households as a percentage of the total[b]	40%	48%	56%	68%[c]
Prime Minister	Macmillan	Heath	Thatcher	Major

[a] Source: *Annual Abstract of Statistics*, various years.

[b] Source: *Social Trends*, 1983, Table 8.6 and 1992, p. 146.

[c] End 1990.

of lone parents has risen from fewer than half a million to one-and-a-third million. As we will see, this big new group now forms a large part of the poorest section of the population. There are also considerably more pensioners than there used to be swelling their impact on inequality. The proportion of owner-occupiers has also risen a great deal, largely at the expense of the private rented sector.

There have also been a whole series of government policy changes that have had a fundamental impact on the inequality of net incomes. The most obvious of these have been the changes in the tax and social security system which are discussed in more detail in Chapter 7. Benefits rising sporadically in the 1960s, though faster than prices overall, and then in line with rises in earnings during the 1970s, along with rising taxes will have had a general dampening effect on inequality. By contrast, the reductions in direct tax rates and failure of benefits to rise with earnings during the 1980s had a disequalising effect.[2]

---

[2] See Johnson and Stark (1989), Johnson and Webb (1993), and Giles and Johnson (1994).

Although no formal incomes policies have been pursued since 1979, it is important to remember that they are one other direct sort of intervention by government that had a major impact on the income distribution during the 1960s and 1970s. Especially where they had a flat-rate element to them, as did most of the policies of the 1970s, these will have had an equalising effect on earnings which, it is worth again stressing, are by a distance the largest component of total income.

Finally, one should recognise the role of Wages Councils, eventually abolished in October 1993, which set minimum wages in a number of industries. Their powers were diminished gradually during the 1980s before their eventual abolition. In placing a floor under wages, they are likely to have reduced what might otherwise have been a greater rise in wage inequality, but their diminishing role will have reduced that impact.[3]

INCOME INEQUALITY

So what has happened to income inequality in this period? Figure 3.2(a) shows what has happened to the Gini coefficient. Four periods can be distinguished from the figure. The first, between 1961 and 1968, saw a very slight fall in inequality. This was followed by a rise between 1968 and 1972 and a 5-year period of falling inequality until a turning-point in 1977. This period coincided with the most active incomes policies of the three decades and benefits increasing in line with the higher of prices and earnings. But interestingly it also coincided with a near doubling in the level of unemployment to around a million in 1977.

The last period, from 1977 onwards, saw a continuous rise in the level of inequality as measured by the Gini, which rose from 0.25 to 0.34 between 1979 and 1993. By comparison with the preceding twenty years or so, this level of change and continuous increase was unprecedented. Indeed, such a change appears to be historically unusual when judged against a longer period.

The changing Gini coefficient itself measures the changes in the shares of income received by the different parts of the distribution.

---

[3] Johnson and Stark (1991) look at the effect of a national minimum wage on overall inequality.

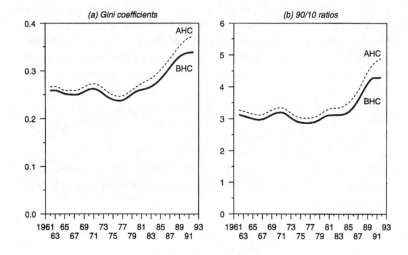

Fig. 3.2. Inequality measured by the Gini coefficient and the '90/10' ratio
Note: Three-year moving averages.

Some idea of how these shares have changed is given in Table 3.2, which shows the income share of the bottom decile, the bottom five deciles, and the top decile in the four periods 1961–63, 1971–73, 1981–83, and 1991–93.

The most startling thing about the first three columns is just how similar they are. The biggest change over this period seems to occur right at the bottom of the distribution, where there is a small growth in the proportion of total income received by the bottom decile. It

Table 3.2. Percentage income shares (BHC) of the bottom decile, bottom half, and top decile

	1961–63	1971–73	1981–83	1991–93
Bottom tenth	3.7	3.9	4.1	2.9
Bottom half	32.6	32.2	32.1	27.1
Top tenth	21.2	21.4	21.3	26.2

actually reached a peak of about 4.4 per cent in the period 1976–78, not shown in the table. But the difference between each of these columns and the column showing the position in the early 1990s is most remarkable. Having received just over a fifth of all income through the 1960s, 1970s, and early 1980s, the richest decile accounted for over a quarter of the total by the early 1990s. At the same time, the share of the lower half of the population fell from just under a third of the total to not much more than a quarter. The share enjoyed by the poorest decile fell to below 3 per cent in 1990 for the first time in thirty years.

Another simple way of looking at inequality is just to look at the ratio between the ninetieth and tenth percentiles of the income distribution (the 90/10 ratio). The changes in this are displayed in Fig. 3.2(b). The line follows the pattern described by the Gini coefficient very closely, rising even faster towards the end of the period, if anything. Like the Gini, the pattern between 1966 and 1978 follows much what one would expect from the incomes policies that were in place at the time.

The steeper rise of the 90/10 ratio towards the end of the period reflects the fact that more was happening at the extremes of the distribution than in the middle. Those at the ninetieth percentile and above did really extremely well from the mid-1980s onwards, while the experience of the very poorest did not reflect any of the growth in incomes experienced by the rest of the population. This pattern is confirmed in Fig. 3.3, which compares the real incomes (expressed in January 1995 prices) of the fifth percentile, the median, and the ninety-fifth percentile. Between 1961 and 1983, the income of the fifth percentile rose from £57 per week to £90 per week (a rise of 58 per cent) which was still its level in 1993. By contrast, the median, which rose from £132 in 1961 to £174 in 1983 (a rise of 32 per cent), rose by a further 27 per cent to £221 by 1993. The ninety-fifth percentile income rose by 40 per cent in the first period, faster than the median but not as fast as the fifth percentile. But in the space of just ten years from 1983, it rose by a further 49 per cent.

If one looks at what happened to incomes after housing costs, the picture is even more stark. For the fifth percentile, the income in 1991 was exactly the same as it was in 1966, and actually about 20 per cent lower than it was in 1979. This partly, of course, reflects growing housing costs. These will not only have depressed the growth of the AHC measure but will actually have inflated the growth of the BHC

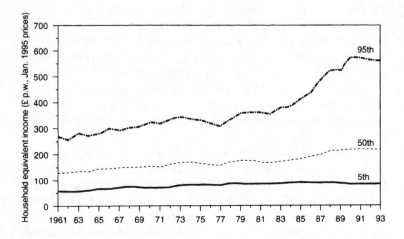

Fig. 3.3. Fifth percentile, median, and ninety-fifth percentile BHC incomes

measure because this measure includes Housing Benefit. If real rents rise, then so will real Housing Benefit payments and hence so will this measure of real incomes for Housing Benefit recipients.

Another way of thinking about these changes is to ask a question such as 'where would the person at the Xth percentile in one year appear in the distribution in another year?'. Let us start by looking at the ninety-fifth percentile in 1961—an equivalent income of £270 per week. By the early 1990s, this income would have placed someone just below the sixty-fifth percentile, though even in 1982 it would have been enough to keep them at the eighty-fifth percentile. Someone in the ninety-fifth percentile in the late 1970s would only be at around the seventy-fifth percentile in the early 1990s. Median income in 1961 would only have placed someone in the second decile in the early 1990s.

THE CHANGING SHAPE OF THE INCOME DISTRIBUTION

As well as looking at an overall measure of income inequality such as the Gini coefficient or the 90/10 ratio, we can gain a number of clear insights into the way that income inequality has changed over the period by looking at the shape of the distribution itself.

In order to focus exclusively on distributional issues rather than on increases in average living standards, we will be examining the distribution of incomes *relative to average income* in the year in question. We use data that are pooled over 3-year periods in order to overcome potential problems associated with the small FES sample sizes in the early 1960s. For reference, Table 3.3 shows how real mean incomes have changed on our two definitions for the four groups of years we will be examining—1961–63, 1971–73, 1981–83, and 1991–93. The significant real increases in the mean, especially after 1981–83, are quite apparent.

Figure 3.4 shows the distribution of incomes on both the BHC and AHC definition for each of our selected groups of years. There are one or two interesting patterns that are picked up on the AHC measure but not on the BHC measure. Whilst it is impossible to summarise the trends over three decades by picking four isolated years, these charts do give an indication of the general trends.

The horizontal axis in each picture gives ranges of income as a proportion of mean income in the year in question. The left-most range covers negative incomes (relevant only on the AHC definition), the next covers incomes up to 10 per cent of the national mean, the next incomes between 10 and 20 per cent, and so on up to the penultimate bar which covers incomes between 340 and 350 per cent of mean income. The final bar includes all incomes in excess of 350 per cent of mean income. The vertical axes are all on the same scale and range from 0 to 7 million individuals.

A number of clear distributional changes can be identified from Fig. 3.4.

Table 3.3. Mean income (BHC and AHC) (January 1995 prices)

	BHC	AHC
1961–63	£146	£131
1971–73	£179	£161
1981–83	£193	£169
1991–93	£266	£232

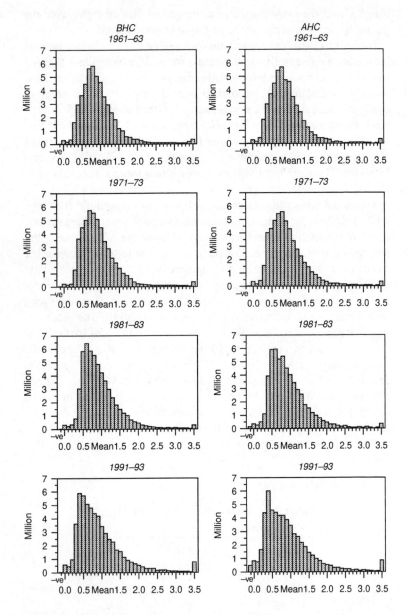

Fig. 3.4.   UK income distribution, BHC and AHC

1. *Shifting peak.* The heaviest concentration of incomes is now to be found at around half the national mean rather than at around 80–90 per cent three decades ago. In 1961–63 (on the BHC measure), the tallest bar was on the 80–90 per cent range where 6 million individuals were to be found, and the second tallest was in the 70–80 per cent range. By 1981–83, the tallest bar was at 60–70 per cent of the mean, with about 6.4 million individuals falling into this income range. By 1991–93, there was a clear peak at 40–50 per cent of the mean with nearly 6 million individuals having incomes in this range. In this latest year, there is also a clear difference between the BHC and AHC pictures. There is a much more obvious peak in the same place on the AHC income measure. This is a peak reflecting benefit levels excluding Housing Benefit which can again obscure part of what is really going on if exclusive consideration is given to the BHC measure.

2. *Lengthening tail.* Over each decade, the tail of the distribution has lengthened. In 1961–63, there were relatively few individuals with household incomes more than twice the national average, whereas by 1991–93 in particular there were several million individuals in this position and more than twice as many with incomes in excess of three times the national average. The whole distribution tails off much more gradually to the right. In 1961–63, there were steep falls either side of the modal income range. By 1991–93, there was a steeper fall to the left but a much more gentle fall to the right. This is the clearest possible illustration of the way in which the distribution has widened.

3. *Proliferation of negative AHC income.* The cluster of individuals below zero AHC income in 1991–93 is much higher than in any other year we have shown. This reflects the high mortgage interest rates that have pushed mortgage holders into negative incomes, and also growing numbers of people reporting losses or zero profits from self-employment.

The pictures of the income distribution effectively showed the numbers of people with incomes at various proportions of the mean. In many studies of relative poverty, the proportion of the population below a particular proportion of the mean is taken as a crude measure of the degree of poverty or of low incomes. Chapter 8 will look at the whole question of poverty in considerable detail, but here we simply note that any such definition of poverty is very much a relative one.

And although this is not the place to concentrate on poverty, one can readily see the proportions of the population below various proportions of the mean by using the numbers on which Fig. 3.4 is based.

Let us take half the mean and the BHC measure of income. In 1961, 10 per cent of individuals had household incomes below this level. That proportion fell to 8 per cent by 1968, rose briefly to 11 per cent in 1972, before falling to a low for the period in 1977. Already it is clear that this measure follows overall inequality trends quite closely. The very fast increase since 1977 and especially over the second half of the 1980s confirms this. By the early 1990s, one-fifth of the population had incomes below half the contemporary mean. One can see the broad outline of this trend quite clearly in Fig. 3.4 just by looking at the way the peak has shifted to below the half-average-income mark over time. On the AHC measure, one-quarter of the population had incomes below half the average by the early 1990s as against 7 per cent in 1977 and 11 per cent in 1961.

This story is mirrored at the other end of the distribution. In 1961–63 and 1971–73, there were around 600 000 individuals with incomes more than three times the contemporary mean. This had fallen back to half a million in 1981–83, but by the end of the period over a million people had incomes over three times the mean.

CHANGES ACCORDING TO FAMILY TYPE

We saw in the last chapter how families of different sorts were concentrated at different parts of the income distribution. In this section, we examine the way in which the positions of the different family types have changed over time. To start, Table 3.4 shows how

Table 3.4. Percentages of population by family type

Family type	1961–63	1971–73	1981–83	1991–93
Couple pensioner	6	8	9	9
Single pensioner	7	7	8	8
Couple with children	45	46	43	37
Couple, no children	22	22	19	23
Single with children	2	3	5	7
Single, no children	17	14	17	16

the composition of the population has changed over time, giving the percentage of the population in each of the six different family types identified in Chapter 2 in the four selected groups of years used in the previous section—1961–63, 1971–73, 1981–83, and 1991–93.

The first trend that is immediately identifiable is the increase in the proportion of pensioners. At the start of the period, married and single pensioners between them accounted for about 13 per cent of the population. This had risen to around 15 per cent at the start of the 1970s, and towards 17.5 per cent in the 1990s. At the start of the period, fewer than one person in seven was a pensioner. By the end of the period, more than one in six were pensioners. Furthermore, the relative positions of single and couple pensioners were reversed. There were slightly more single pensioners at the start of the period, but significantly more married ones by the end.

Individually, the non-pensioner groups follow much less of a steady trend. Couples with children formed about 45 per cent of the population for the first twenty years, but this proportion dropped off quite dramatically over the 1980s such that they formed just 37 per cent of the population by the early 1990s. The biggest proportionate change was among single-parent families. Individuals in such families formed just under 2.5 per cent of the population in 1961–63, more than doubling their representation to over 7 per cent of the population by the 1990s.

The reason for having spent some time describing how the family-type make-up of the population changed is that it is necessary to keep that in mind when considering to what extent each family type changed its over- or under-representation at any point in the income distribution. Clearly a doubling of the number of members of single-parent families at any point in the income distribution is only to be expected on the basis of the doubling of their representation in the population as a whole. On the other hand, it is worth bearing in mind the findings of Goodman, Johnson, and Webb (1994) that few of the changes in the overall income distribution can be explained by reference to changes in the demographic structure of the population.

Given that we have thirty-three years of data, it is not going to be possible to present detailed results on the demographic make-up of the entire income distribution for each year. Instead, attention is devoted to the top and bottom ends of the distribution, for this is where most of the interesting changes have occurred. And although we do go on in this chapter to look at the income distributions among pensioners and non-pensioners separately, the greater part of the description of the degrees of inequality within and between different family types is reserved for Chapter 6, in which explanations for the changes in overall inequality are sought.

Figure 3.5 shows graphically how the composition of the bottom decile has changed over time. A number of trends are identifiable. Back in the early 1960s, pensioners made up over 40 per cent of this poorest group. By 1991, this proportion had fallen to barely 20 per cent. The main groups to replace these pensioners in the bottom decile group have been the single childless and couples with children. Most of this change has in fact occurred since the mid-1970s, when the large increases in unemployment occurred, indicating that part of what has happened is that the unemployed below pension age have pushed those over pension age out of this poorest group. Prior to the mid-1970s, the structure of the bottom decile group was quite stable.

Looking back at the trends described in Table 3.4, it is evident that, for example, while the proportion of the poorest decile who are

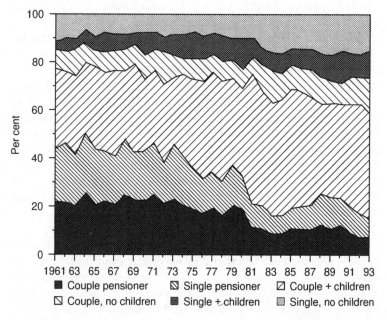

Fig. 3.5. Composition of the bottom decile group by family type

pensioners has fallen, pensioners have been increasing their representation in the population as a whole. So to that extent Fig. 3.5 understates the change in the relative over-representation of pensioners in the poorest group. In fact, back in the early 1960s, the proportion of pensioners in the poorest decile group was over three times the proportion of pensioners in the population as a whole. By the end of the period, couple pensioners were not over-represented at all in the bottom decile. In other words, the proportion of the poorest decile made up of couple pensioners was little different from the proportion of the population as a whole who were couple pensioners. Single pensioners were over-represented by a third as opposed to by three times in 1961.

In the 1990s, single-parent families are the most heavily over-represented at the bottom of the income distribution, being nearly one-and-a-half times as common in the poorest decile as in the population as a whole. This level of over-representation is only half that of pensioners at the start of the period. There has been a general convergence of all groups towards equal representation at the bottom of the income distribution and in the population as a whole. In general, family type is a much less decisive determinant of the likelihood of being in the poorest decile than once it was.

Even couples with children are now over-represented at the bottom. They have always formed a large part of the poorest decile because they represent such a large section of the population, but whereas they used to form a smaller part of the bottom decile than of the population as a whole, since about 1980 they have actually formed a larger part of this group than of the whole population. Childless non-pensioners, whether single or couples, have been under-represented in the bottom decile throughout.

Changes in the bottom quintile have been less dramatic. Within this group, there has been only a modest improvement over time in the relative position of pensioners. From constituting a third of the bottom quintile in 1961, pensioner representation there rose to almost 40 per cent in the early 1970s before falling back to just under a quarter by the end of the period. The change in the representation of single-parent families has, by contrast, been greater here than in the bottom decile. From constituting just 5 per cent of the bottom quintile in 1961, single-parent families made up 15 per cent of this quintile in the early 1990s. Many pensioners and single parents have just enough money to stay out of the poorest decile, but not enough to escape from the poorest quintile. Table 3.5 shows the breakdown of the bottom quintile by family type for selected years.

Once again, a slightly different picture emerges if the after-housing-costs income measure is used for looking at the composition of the poorest groups. For the bottom decile, the trends are even more striking. The AHC measure tends to improve the relative position of those who own their home outright or who have paid off most of their mortgage, and over the period, increasing numbers of pensioners came to fit this description and so were lifted out of the poorest group. Those with mortgages, by contrast, did relatively worse as the period drew to an end. So while pensioners' share of the bottom decile fell from over 40 per cent to around 20 per cent on the BHC measure, there was a fall from a similar starting-point to less than 10 per cent on the AHC measure. Conversely, each of the non-pensioner groups saw a steeper rise. Again, the patterns are far less clear when the bottom quintile is considered.

Moving to the other end of the distribution, we know already that the top decile income rose very much faster than that at other parts of the distribution. Nevertheless, the family-type composition of the richest decile changed very little over time. Through the whole period, couples with and without children and single people accounted for around 90 per cent of the richest tenth, with pensioners more or less steady at around 10 per cent and almost no lone parents. Over the whole period, relative to their population share, childless couples were the most over-represented in this richest group. There have been few trends to describe.

Table 3.5. Composition of bottom quintile by family type

Family type	1961–63	1971–73	1981–83	1991–93
Couple pensioner	17	20	15	13
Single pensioner	17	19	12	10
Couple with children	43	36	43	37
Couple, no children	8	9	8	11
Single with children	5	8	9	15
Single, no children	10	8	13	14

PENSIONERS AND NON-PENSIONERS

As well as being concerned with the composition of the richest and poorest groups, we are concerned with the distribution of income within various family types and trends in their real income levels. More detail on this is provided in Chapter 6, but here we provide some evidence on the different experience of pensioners and non-pensioners. In terms of their average income levels, Fig. 3.6 shows how little the relativities changed over the period 1961–93. Average pensioner income was around three-quarters of average non-pensioner income at the start of the period, as it was at the end of the period, and deviated little in the intervening years. Given the different forces shaping the incomes of the two groups, this stability is remarkable. For the incomes of non-pensioners are shaped by earnings levels, unemployment levels, and labour-force participation. Pensioners' incomes are shaped by these things *in the past*, but by current social security policies and interest rates in the present. The two periods in which there was some small closing of the gap between pensioners and non-pensioners were in the recessions of the mid-1970s and early

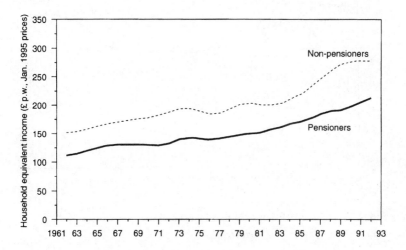

Fig. 3.6. Mean incomes of pensioners and non-pensioners
Note: Three-year moving averages.

1980s when average non-pensioner incomes actually fell whereas the incomes of pensioners were to some extent protected by the social security system.

In fact, this relative stability in the relationship between the average incomes of pensioners and non-pensioners hides two quite different experiences in the development of inequality among the two groups. Figure 3.7 shows the development of the Gini coefficient for pensioners and for non-pensioners separately. The patterns are quite different. The pensioner Gini started off significantly above the non-pensioner Gini at about 0.3 as against 0.25, but fell gradually and fairly continually before flattening at the end of the 1970s at a value of 0.22 and only starting to rise again in about 1983. By the end of the 1970s, the pensioner Gini had dipped under that of the rest of the population and has remained there, though it did rise again swiftly from the mid-1980s,[4] again approaching 0.3 at the end of the period.

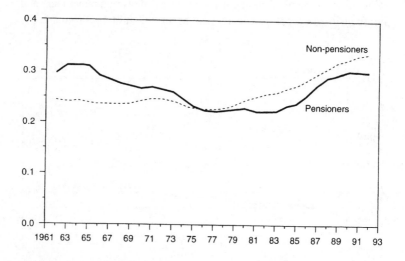

Fig. 3.7. Gini coefficients for pensioners and non-pensioners
Note: Three-year moving averages.

---

[4] The rise in inequality among pensioners in the 1980s is described in more detail by Hancock and Weir (1994) and Dilnot and Johnson (1992). The trend over the whole period is explored by Johnson and Stears (1995).

The initial fall in pensioner inequality can be put down largely to falling employment participation among those over state pension age and rising social security benefits. The later rise reflects stable benefits and growing private pensions and investments for some.

The pattern of the Gini for non-pensioners is much more like that of the population as a whole.

CHANGES BY ECONOMIC STATUS

Again, we start by looking at how the population shares of the different economic status groups have changed over time. This is illustrated in Fig. 3.8. A number of clear trends are discernible. There has, of course, been a rise in unemployment over the period. In our data, individuals living in families containing only the unemployed made up only about 1 per cent of the population through the 1960s, reaching above 2 per cent for the first time in 1972 and really taking

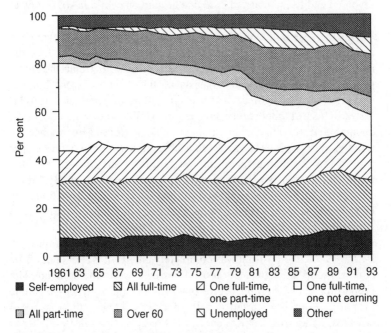

Fig. 3.8. Composition of the population by economic status

off from 1979 onwards to reach a peak of 8 per cent in 1987, before falling back to 6 per cent at the end of the period.

Even more noticeable is the remarkable diminution in the proportion of people living in families with one earner and one non-earner— the traditional single-earner couple, often with children. In the early 1960s, over a third of people lived in single-earner couple families. This proportion had more than halved by the end of the period. This stereotype of the typical family of working husband, non-working mother, and children has become increasingly unrealistic as a number of factors have contributed to its demise. First, there has, of course, been the decline in male full-time employment: between 1971 and 1991, the proportion of working-age men in full-time employment fell by almost a fifth. Second, there has been a massive rise in female part-time employment: whilst over the last two decades there has been very little change in the proportion of working-age women in full-time employment, the proportion in part-time work has risen from around one in five in the early 1970s to one in three by the early 1990s. Finally, of course, there has been the rise in the number of divorced and never-married mothers: around one child in six is now to be found in a lone-parent family.

This increase in the number of lone-parent families has in fact been partly responsible for the other major change visible in Fig. 3.8—the doubling of the numbers in the 'other' category. Other than the single parents, this group largely consists of those out of work but not seeking work for a variety of reasons including ill health and straight-forward loss of hope of finding a job. Unlike the numbers in the unemployed category, the numbers in this group rose unchecked through the 1980s. Having been stable at between 4.5 and 5 per cent of the population for twenty years up to 1981, the share of the population in this group had risen to 8.5 per cent by the early 1990s.

Looking at what has happened in the bottom decile over time, the hypothesis that pensioners have been pushed out by the unemployed is strongly supported. In 1961, 44 per cent of those in the bottom decile were in the over-60 category, with only 4 per cent being unemployed. By the 1990s, the position was completely changed: 24 per cent of the bottom group were over 60 and 30 per cent were unemployed. The proportion of the bottom decile that was unemployed was unstable over the 1960s but started a path of consistent growth from 1974 onwards, rising from 10 per cent in 1974 to 20 per cent in 1977, 32 per cent in 1981, and a high of 36 per cent of the bottom decile in 1982. It

fell back steadily to 21 per cent in 1990 but jumped to 30 per cent the
next year as the recession of the early 1990s resulted in very swift
rises in unemployment. The other interesting pattern at the bottom
relates to the self-employed who, over the 1980s, made up an increasing proportion of the poorest group. During the 1970s, the self-
employed averaged just under 9 per cent of the bottom decile and
this was also true of the first half of the 1980s. But from 1986 on, they
averaged just under 14 per cent of the group. This reflects an increasing number of low-skilled unemployed individuals trying to return to
the labour market, being unable to find a job, and using self-employment as a way in. It was also possible to take advantage of government encouragement through programmes such as the Enterprise
Allowance Scheme.

With the exception of the 'other' category, representation of the
other groups has been small throughout the period. In Table 3.6,
therefore, we include only the self-employed, the over-60s, the unemployed, and the 'other' group.

Again, the picture for the bottom quintile is somewhat different,
with less dramatic swings in the fortunes of the different groups. The
proportion of the bottom quintile taken up by the over-60s was very
similar at the end of the period to the proportion at the start of the
period, though the share of the group who were unemployed rose from
between 2 and 5 per cent in the 1960s to 21 per cent by 1991, having
reached a high of 28 per cent in 1987. The group that the unemployed
replaced in the bottom quintile were a group who at no point made up a
significant part of the bottom decile but who in the 1960s comprised a
quarter of the bottom quintile. By the end of the period, they accounted
for just 8 per cent of this quintile. They are the single-earner couples—
one in full-time work, one not working. Partly this massive fall reflects
their diminished share in the population as a whole, partly it reflects the

Table 3.6. Composition of bottom decile by economic status
(Percentage in selected groups)

Economic status	1961–63	1971–73	1981–83	1991–93
Self-employed	8	7	10	15
Over 60	47	48	23	20
Unemployed	6	13	34	31
Other	21	17	14	18

fact that families with a full-time worker were not only well off enough by the end of the period to escape the bottom decile but to escape the bottom quintile as well. Those out of work became both more numerous and relatively poorer, so displacing them.

As we saw in Fig. 3.8, there have been large changes in the economic status composition of the population as a whole, so the compositional changes in the bottom decile and bottom quintile are potentially consistent with no change in the relative likelihood of a member of any particular group falling into this income band. In fact, there have been changes in relative likelihoods of being in the bottom income bands, but not changes that one would necessarily have expected from the discussion so far.

Perhaps most surprisingly, there has in fact been a fall in the over-representation of the unemployed in the poorest decile. Although they made up a much greater proportion of the poorest group at the end of the period, their over-representation relative to their total numbers actually fell gradually from about 6 times to 4.3 times by the late 1980s, before climbing again to 4.7 times in the recession of the early 1990s. By 1993, the over-representation of the unemployed had fallen back again to about 4.2 times. While the chances of the poorest being unemployed rose dramatically, the chances that any unemployed person would be among the poorest actually fell. There was an even more significant fall among the 'other' group and, as expected, among the over-60s. The main group to see their relative likelihood of being in the bottom decile rise was the self-employed who, after spending most of the 1960s being slightly under-represented in the poorest decile, were one-and-a-half times over-represented by the end of the 1980s.

These results are also sensitive to the choice of decile rather than quintile. There was no significant fall in the over-representation of the unemployed in the bottom quintile, where they were roughly three-and-a-half times over-represented throughout the period.

At other parts of the distribution, the changes in composition have been much less marked. Right at the top, full-time workers and the self-employed made up about 60 per cent of the tenth decile throughout the period. The self-employed actually reduced their level of over-representation from over two times to around one-and-a-half times.

There have also been considerable changes over the period in the distribution of incomes within the various economic groups. Most of the discussion of these changes is reserved for Chapter 6, in which the role of changing within-group inequalities in the overall changing income distribution is discussed. Here, it is worth making one or two pertinent observations, and Fig. 3.9 shows the trends in inequality within economic status groups, as measured by the Gini coefficient. All those whose family unit contains someone in full-time employment are classified as 'full-timers', whilst those containing only part-time workers (a small fraction of the population) have been omitted from the figure.

The incomes of the self-employed have been quite diverse throughout the period. They are one of the few groups to have been reporting significant numbers of very high incomes in the 1960s and 1970s. But during the 1980s, there was, as we have already noted, a change in the nature of self-employment, with a marked increase in the number of self-employed reporting losses or nil profits.

There were fewer families containing a full-time worker later on in the period, but the group performed well relative to the national average, with many more receiving incomes well in excess of the

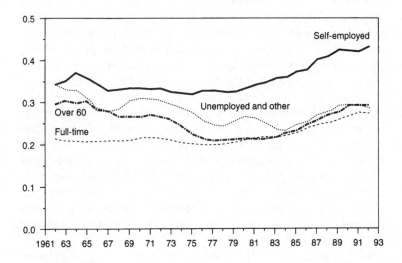

Fig. 3.9. Gini coefficient for economic status groups

Note: Three-year moving averages.

national average than was the case in earlier years. Finally, we can say something about the economically inactive—the unemployed and 'other' categories—whose numbers doubled. The incomes of this group are much more concentrated than those of workers because benefits form such an important element of their incomes. The unemployed have also enjoyed much less in the way of real income increases than those in work. More than 2 million of them had household incomes between 30 and 40 per cent of the national average by the early 1990s.

REGIONAL CHANGES

The regional composition of the population has changed by much less than the family or economic status compositions. There has been a small shift in the population away from the North and towards the South, but this has been a matter of no more than a couple of percentage points over a period of thirty years. But the gap between the richer and poorer regions would appear to have grown.

Back in the early 1960s, mean incomes in the North were roughly 90 per cent of those in the South (including London). By the end of the 1980s, incomes in the North were nearer three-quarters of those in the South. The Midlands also fell well behind the South, having started with incomes very similar to those in the South and certainly within 5 per cent of them. By the end of the period, average incomes in the Midlands were about 80 per cent of those in the South.

Wales and Scotland, which started with very similar average incomes—about 5 per cent below those of the North—have since diverged. Average incomes in Scotland ended up very close to those in the Midlands and higher than those in both Wales and the North. Scottish incomes appear to have risen ahead of Welsh incomes in the 1970s, to have returned to near parity in the recession of the very early 1980s, but to have moved gradually further ahead from the mid-1980s onwards. These trends are all illustrated in Fig. 3.10, which shows the average incomes in each region.

Just as one would expect, the poorer regions tend to be over-represented at the bottom of the income distribution and the richer regions under-represented. Under-representation of the Southern English in the poorest decile became more marked over time,

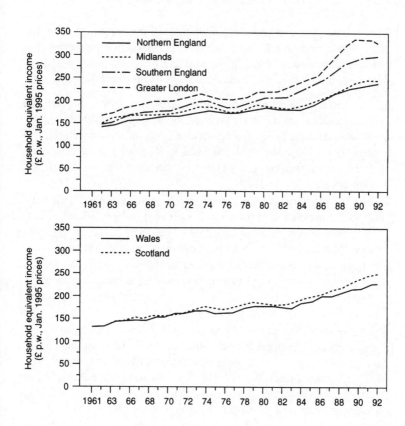

Fig. 3.10. Mean incomes by region

Notes: Three-year moving averages. There are no data points for 1964.

particularly between the end of the 1970s and mid-1980s. Scotland's over-representation at the bottom followed the trend seen in average incomes, falling particularly towards the end of the 1970s. There is a gradual growth in the over-representation of Northerners through to the mid-1980s and then a steadying. The Midlands, slightly under-represented at the start of the period, suffered a large increase in numbers in the bottom decile during the middle of the 1980s.

SUMMARY AND CONCLUSIONS

This chapter has concentrated on changes in the income distribution since 1961, over which period a consistent definition of income has allowed us to look in detail at the numerous changes that took place. The main points that arise from this analysis are the following:

1. The increase in income inequality during the 1980s dwarfed the fluctuations in inequality seen in previous decades. Whilst inequality declined gradually during much of the 1960s, rose slightly to the early 1970s, and then fell back to its lowest point in around 1977, these fluctuations were very modest compared with the changes seen in the 1980s.

2. The income share of the poorest tenth of society has fallen back from 3.7 per cent in 1961–63 to 2.9 per cent in 1991–93, with most of the fall occurring during the 1980s. The share of the richest tenth rose from 21 per cent to 26 per cent over the three decades.

3. Whilst real incomes (before housing costs) have grown by around 84 per cent on average over the last three decades, the incomes of the richest tenth have risen twice as fast (up 109 per cent) as those of the poorest tenth (56 per cent).

4. Taking into account the effects of housing costs can greatly affect assessment of changes in real living standards. The real incomes of the poorest tenth ranked by income after housing costs actually fell sharply from a peak in 1979 of £75 per week to just under £60 per week in 1993 (both in 1995 prices). This represented a return to the living standards of more than a quarter of a century ago.

5. In terms of the composition of the poorest groups, a major change has been the relative improvement of the position of pensioners. Pensioners formed almost half of the poorest decile group in 1961, compared with less than a quarter in 1993, and this despite a significant growth in the number of pensioners in the population. However, pensioners are still among the poorer groups, with their representation in the poorest quintile down much less sharply.

6. The emergence of mass unemployment has had a major effect on the income distribution. Families with children now make up more than half of the poorest decile group compared with only around a third three decades ago, with the main reason for this change being the more than eightfold increase in unemployment between the early 1960s and the mid-1980s.

7. The self-employed have become an increasingly important group in the last ten to fifteen years, and the outcomes of the self-employed have become more diverse. The self-employed are the only economic group in the 1980s to be systematically over-represented both at the very bottom *and* at the very top of the income distribution.

4 The Distribution of Expenditure

INTRODUCTION

The last two chapters have looked at the distribution of income and how it has changed over the last thirty-three years. This chapter compares the changes described with a different measure of living standards, household expenditure.

There is no single correct way to measure living standards and, as was seen in Chapter 1, the measurement choices that are made matter for the picture of living standards drawn. For example, inequality of expenditure has not risen as fast over the last fifteen years as inequality of income, and many of those who would be classed as amongst the poorest by their income are different people from those who would be counted amongst the poorest by their spending.

Although these differences, and others to be described, indicate that we should be cautious about making firm conclusions about living standards from any one measure taken in isolation, the different pictures of living standards drawn by looking at household income and household spending do not provide competing views. Taken together, a good deal more can be discovered about the nature of inequality and living standards than by using any single measure on its own.

This is because different measures of living standards pick up different aspects of people's well-being. The measure of current income that has been used in the analysis thus far provides a snapshot of incomes across the population at a given point in time. One draw-back of this measure is that, in some cases, it may not be a good reflection of well-being over a longer period. If people's incomes vary over time, then the population is likely to contain some people whose current incomes are unusually high or unusually low, and so do not fully reflect the amount of resources available to them over the longer run.

Two approaches are used in this book to widen the perspective from just the 'snapshot view' of living standards. The expenditure measure presented in this chapter provides a longer-run view of living standards since, if people are able to, they base their expenditure decisions not just on their income at a point in time but on their view of how well off they expect themselves to be over the longer term. Chapter 9 takes another approach by considering the dynamics of income, looking at, for example, the extent to which individuals' incomes fluctuate from year to year. Together with what we know about current incomes, both of these approaches can tell us more about living standards than looking at current income alone.

This chapter examines the main features of the distribution of expenditure and how it has changed over time. It points out the main similarities to and differences from the changes in the income distribution that were described in Chapters 2 and 3, and considers some possible ways in which these differences can be explained.

In this chapter, we concentrate on trends in spending and incomes over the period 1968–93, since expenditure data that are directly comparable to the income data we have used are not available for the earlier part of the 1960s. In the last two chapters, we have presented results for income on both the before- and after-housing-costs measure. In this chapter, we focus only on the income measure before housing costs have been paid for, and compare this with household expenditure that includes spending on housing. Comparisons of after-housing-costs income and expenditure reveal very similar results, and so are not set out here.[1]

INEQUALITY

Different households in the UK have very different levels of weekly spending. Compared with median spending of £211 (expressed in terms of the equivalent level for a childless couple) in 1991–93, the top tenth of households spent more than £450 per week whereas the lowest tenth spent less than £110. In fact, the level of inequality of expenditure between households in 1991–93 was roughly the same as, or very slightly lower than, the level of inequality in the distribution

---

[1] Goodman and Webb (1995) provide some results on an AHC basis for household expenditure and income.

of income, depending on which measure of inequality is used. The household at the ninetieth percentile of expenditure spent 4.1 times more than the one at the tenth percentile of expenditure, whereas the 90/10 ratio for income was 4.3. The Gini coefficient in 1991–93 was approximately 0.34 for both income and expenditure.

Although the level of inequality in the two distributions is now about the same, comparing the *level* of inequality in income and spending is less informative than comparing the *changes over time* that have taken place. This is because of the measurement problem arising from the 'lumpiness' of many items of expenditure that we discussed in Chapter 1. As we saw in Chapter 1, some of the measured inequality in spending between different households arises just from differences in the timing of purchases. This is not the sort of inequality that is interesting for the comparison of living standards.

But so long as the timing of households' purchases has not changed very much over time, and this is a serious issue, then although the level of measured inequality might not tell us much, *changes* in the inequality of spending, and how these compare with the changes in income inequality over time, can be important and informative.

The trends in the distribution of income and expenditure over time have indeed been very different. Looking at inequality of the two distributions back to 1968, Fig. 4.1 shows the Gini coefficients and the 90/10 ratios of income and expenditure in each year. It shows that up until the late 1980s, expenditure was always more unequal than income. Although both income and expenditure inequality grew over the 1980s, income inequality grew considerably faster, and so 'caught up' with the inequality in the distribution of expenditure.

As we saw in the last chapter, underlying the rapid growth in income inequality, over the last decade in particular, has been much faster income growth at the top of the income distribution than at the middle of the distribution, where in turn growth has been much faster than real income growth at the bottom of the income distribution. The changes in the expenditure levels at different parts of the distribution of expenditure are far less diverse. In particular, the lower part of the expenditure distribution has seen spending growth that is much closer to the spending growth of the population nearer the middle of the distribution. Whilst the income of the lowest income decile group has changed little since the early 1980s (or has fallen if income is measured after housing costs), the expenditure of the lowest tenth of spenders has grown over this period.

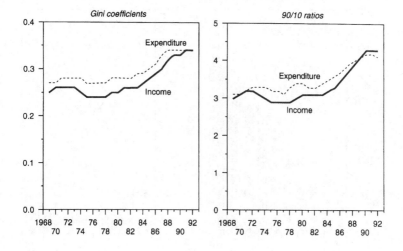

Fig. 4.1. Gini coefficients and 90/10 ratios
Note: Three-year moving averages.

This can be seen in Fig. 4.2(a), which compares the percentage change in income and in expenditure for successive decile groups of the population between 1981–83 and 1991–93. It is important to be clear on the interpretation of this figure: the percentage changes in expenditure show how typical (median) spending within each expenditure decile group has changed over the 10-year period, whilst the income changes show how typical (median) income has changed in each income decile group. From this figure, it can be inferred that if expenditure is used to measure living standards, the bottom decile group's living standards grew by 13 per cent in real terms over the 10-year period; if income is the preferred measure, then living standards in the bottom decile group barely rose at all. Spending growth amongst the second, third, and fourth expenditure decile groups was also higher than income growth amongst these income decile groups.

On the other hand, at the top, it is the income measure of living standards that shows the largest gains over the 10-year period. The income of the top income decile group grew by 54 per cent, whereas

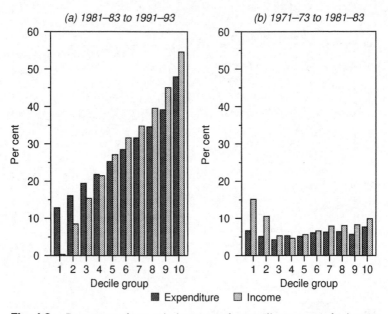

Fig. 4.2. Percentage changes in income and expenditure across the income and expenditure distributions

the spending of the top expenditure decile group grew less, by about 48 per cent.

The changes in income and spending in the 10-year period previous to this are much smaller, but for almost all decile groups, the gains in living standards are higher when income rather than spending is used. Figure 4.2(b) shows percentage changes across the distributions between 1971–73 and 1981–83. This figure is drawn to the same scale as Fig. 4.2(a), and the first thing that is apparent is that the changes in this decade were much smaller, across both the income and the expenditure distributions. Particularly at the lower end of the scales, the growth in income is higher than that in spending. The income of the bottom income decile group rose by 15 per cent over the 10-year period, whereas the spending of the bottom spending decile group rose by only 7 per cent. The income growth at the bottom of the income distribution over this 10-year period was higher than for any other income group across the distribution.

FAMILY TYPES AND ECONOMIC STATUS

When expenditure is used as a measure of living standards rather than income, different sorts of people predominate amongst the poorest groups. One of the most notable features of the changes in incomes over the last three decades has been the displacement of pensioners at the bottom of the income distribution by families of working age. There are also now more families of working age amongst the lowest-spending, but their representation amongst the lowest spenders has not increased to nearly the same extent as when income is used to measure their living standards.

Figure 4.3(a) shows the composition of the bottom income decile group in terms of family types, and Fig. 4.3(b) shows how the composition of the bottom spending decile group compares. Figures 4.4(a) and 4.4(b) show the same decile groups broken down into economic status groups. As we saw in Chapter 3, the number of non-pensioners,

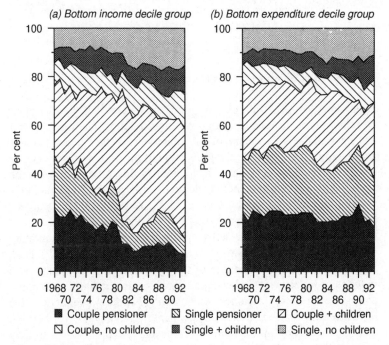

Fig. 4.3. Breakdown by family type

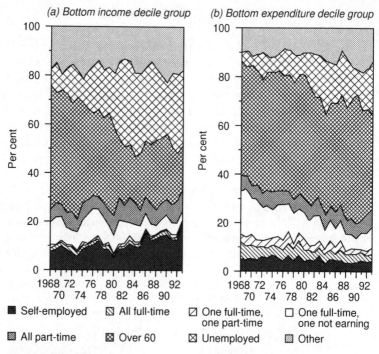

Fig. 4.4. Breakdown by economic status

particularly couples with children and, to some extent, lone parents, on relatively low incomes has grown considerably. The result of this has been the displacement of pensioners from the lowest income decile group. By 1991–93, the proportion of the bottom income decile group made up of pensioners had fallen to only about 17 per cent, from the 40 per cent that they made up over the 1960s. Looking at the population broken down by labour market status, Chapter 3 showed that the over-60s at the very bottom of the income distribution have been partly replaced by the self-employed, the unemployed, and other economically inactive groups such as the long-term sick and disabled.

By contrast, the composition of the lowest tenth of spenders, both in terms of family type and economic status, has remained much more constant over time. Pensioners accounted for about 40 per cent of the lowest tenth of spenders broken down into family types in 1991–93, and this was not much reduced from the 50 per cent that they accounted for at the start of the 1970s. Whilst pensioners are now barely over-represented in the bottom income decile group as compared to their representation in the population as a whole, they are still over-represented amongst the lowest spenders.

There has been some increase in the numbers of people below pension age at the bottom of the expenditure distribution: the numbers of lone parents, and couples with and without children, at the bottom end of the spending scale have risen somewhat, as have the numbers of unemployed and the 'other' economically inactive. The number of self-employed amongst the lowest spending has not grown, remaining at about 4 to 6 per cent of the bottom spending decile group for the whole of the period in question. The self-employed are one group who are over-represented at the bottom of the income distribution and under-represented at the bottom of the expenditure distribution.

The composition of the lowest quintile of spenders has also remained relatively constant over time. Pensioners take up a smaller proportion of this group than of the lowest tenth, showing that those pensioners who are to be found amongst the lowest fifth of spenders in the population are more heavily clustered in the lowest tenth than in the second decile group. This provides another contrast with the distribution of income, where pensioners, particularly couple pensioners, take up a larger part of the second decile group than of the lowest income decile group.

The top decile groups by income and by spending have a more similar composition, and have seen similar changes over time. In Chapter 3, we saw that there have been only a small number of pensioners, and almost no lone parents at all, amongst the top decile group by income; the same is true amongst the top tenth of spenders. By economic status, the same pattern for income and spending is also apparent: those living in family units in which there is a full-time worker predominate amongst the top of both distributions.

Pensioners and Non-Pensioners

In Chapter 3, we saw that, on average, pensioners have lower incomes than non-pensioners, but that the ratio of mean pensioner income to mean non-pensioner income had remained roughly stable (at about three-quarters) over time. Pensioners also have lower spending on average than the non-pensioner population. We have already been given some clue of this by the fact that there are many pensioners clustered in the bottom quintile of the spending distribution, and in

particular in the lowest tenth. But, of course, the mean is determined
by the spending of pensioners across the distribution, not just those at
the bottom.

Figure 4.5 shows how the mean pensioner expenditure compares
with mean non-pensioner expenditure, and should be compared to Fig.
3.6. The figures are remarkably similar: again, relativities have changed remarkably little, although, at around 70 per cent for much of the
period, the ratio between average pensioner and non-pensioner spending is lower than the ratio between average incomes. Again, as we saw
in Chapter 3 with incomes, non-pensioner spending has tended to
fluctuate between times of economic boom and recession more than
has pensioner spending.

The patterns in pensioner and non-pensioner spending inequality
are rather different, although there are fewer similarities between
trends in income and spending here. Figure 4.6 shows pensioner
and non-pensioner Gini coefficients over the period 1968–93. The
non-pensioner Gini coefficient is shaped very much like the Gini
coefficient for the population as a whole, reflecting the larger

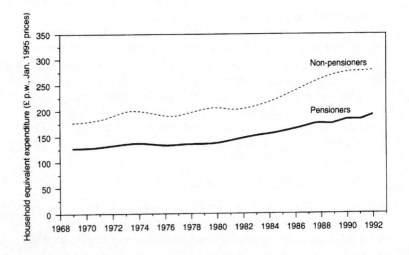

Fig. 4.5.   Mean pensioner and non-pensioner expenditure

Note: Three-year moving averages.

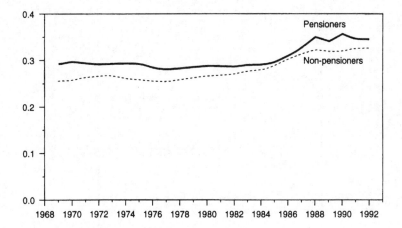

Fig. 4.6. Mean pensioner and non-pensioner Gini coefficients
Note: Three-year moving averages.

population share of non-pensioners. For pensioners, the pattern of inequality does not follow the same well-defined U shape that we saw in the pensioner income distribution. This is because pensioner expenditure inequality did not fall over the late 1960s and 1970s to the same extent as pensioner income inequality did.

INCOME AND EXPENDITURE

There are many individuals whose relative position in the distribution of expenditure is quite different from their relative position in the distribution of income. Table 4.1 shows where members of each income decile group were to be found in the distribution of expenditure in 1991–93. Reading along the rows of Table 4.1, we can see that about one-third of the bottom income decile group are also in the bottom expenditure decile group. The bulk of the remaining two-thirds of them are found amongst the lower half of spenders. A considerable number, almost a fifth of the lowest income decile group, are in the top half of spenders.

Table 4.1. Expenditure vs. income deciles, 1991–93

Decile group	Expenditure: 1	2	3	4	5	6	7	8	9	10	
Income: 1	**34**	19	12	8	7	5	5	4	2	3	100
2	29	**27**	18	10	6	4	2	1	1	1	100
3	17	22	**20**	14	11	6	4	3	2	2	100
4	8	14	17	**19**	14	11	8	5	4	2	100
5	4	7	12	17	**18**	15	11	7	6	3	100
6	3	4	9	13	14	**17**	15	11	8	5	100
7	2	4	5	10	13	18	**18**	15	11	6	100
8	1	2	4	6	9	13	18	**20**	16	10	100
9	1	1	2	3	5	8	13	21	**26**	20	100
10	1	1	1	1	2	4	6	12	23	**47**	100
	100	100	100	100	100	100	100	100	100	100	

Amongst the higher income groups, a large proportion of the very richest by income are also the very richest by spending. Almost half of the richest income decile group are also in the highest spending decile group, whilst less than a tenth of them are in the lowest half of spenders.

Reading down the columns of Table 4.1 shows how households at different levels of spending are ranked according to their incomes. Again, we see that about one-third of the lowest expenditure decile group are also in the lowest income decile group. But the rest of the lowest spenders are not spread so widely throughout the whole of the income distribution in the way that the lowest income group was spread across the expenditure distribution. Almost a third of this group again are to be found just with slightly higher incomes, in the second income decile group. At the top end, the highest spenders are concentrated at the top of the income distribution.

Although, as Table 4.1 shows, there are many who are ranked differently according to their expenditure than according to their income, except at the very bottom of the income distribution, the underlying pattern is one of spending broadly rising with income. Figure 4.7 shows levels of spending within each *income* decile group in 1991–93. The population in each income decile group has been ranked according to its spending, and the spending of a low (twenty-fifth percentile), middle (fiftieth percentile), and high (seventy-fifth percentile) spender within each decile group has been graphed. Across almost the entire distribution, the lines are upward-sloping.

It is only the bottom income decile group whose spending, compared with the income decile group directly above, does not appear to match the general upward-sloping pattern amongst the population. In 1991–93, the highest-spending of the bottom income decile group spent more than those in the income group directly above. This reflects the fact, drawn from Table 4.1, that amongst those with the lowest incomes, there are a significant number whose spending is not ranked amongst the lowest in the population. This is partly because of the number of relatively low-spending pensioners who have escaped the bottom income decile group. It is also because there are some who appear to have very low incomes whose spending is high relative not just to low-spending pensioners but also to the spending of the population as a whole.

This 'tick shape' is only a feature of the income and expenditure distributions over the 1980s and early 1990s. Prior to 1981–83, the

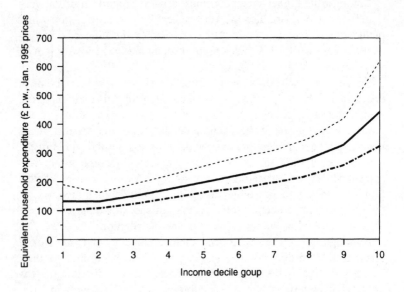

Fig. 4.7. Quartiles of expenditure within decile groups of income, 1991–93

expenditure quartiles rise smoothly with income. This can be seen in Fig. 4.8: the graphs for 1971–73 and 1976–78 show expenditure sloping upwards across the entire income distribution, but the graphs for 1981–83 and 1986–88 show the same shape as in Fig. 4.7, although it is less pronounced. This means that over this period, there has been an increase in the numbers of low-income households that show relatively higher spending. In fact, although there was no overall real change in the *income* of the lowest income tenth between 1981–83 and 1991–93, the *spending* of this lowest income group was about 20 per cent higher at the end of the 10-year period than at the start.

We saw in Table 4.1 that although many of those with the lowest incomes were spread right across the distribution of spending, the converse was not true: the lowest spenders tended to be more concentrated towards the bottom end of the income distribution. This pattern is again reflected in Fig. 4.9, which shows levels of income across the expenditure distribution. Within each expenditure decile group, a low-, middle-, and top-income household has been plotted. In contrast to Fig. 4.7, this graph contains no 'tick shape' at all. Across the entire population, the pattern is one of income broadly rising with

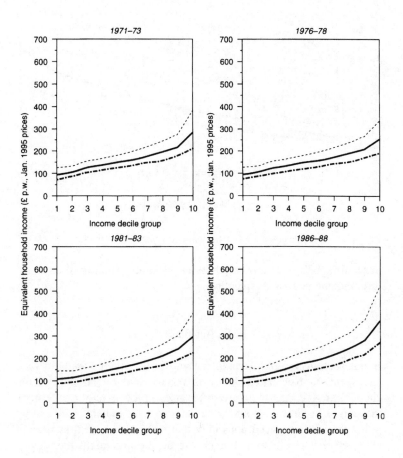

Fig. 4.8. Quartiles of expenditure within decile groups of income, 1971–73 to 1986–88

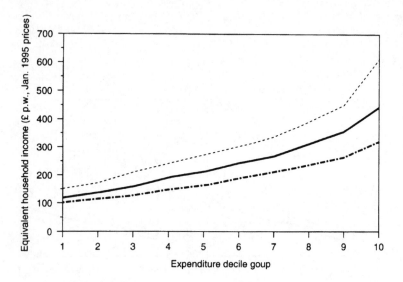

Fig. 4.9. Quartiles of income within decile groups of expenditure, 1991–93

expenditure. Even at the very bottom of the distribution of expenditure, income rises with spending.

EXPLANATIONS

In this section, we look for explanations of why the trends in income and spending have been different. As we said at the start of the chapter, the fact that there are differences in the sorts of conclusions that might be drawn about living standards by looking at spending rather than income need not imply that one of the measures is right and the other one wrong. Rather, the differences might themselves provide information about how living standards have changed.

Income Risk

One important explanation that has been put forward for why income inequality rose faster than spending inequality, particularly over the later part of the 1980s, has been that people's incomes have become more volatile and *uncertain* over time. This could have come about if, for example, more people have become reliant on income from irregular sources, such as self-employment income, or if there is higher job turnover and more people experience spells of unemployment as a result.

One study that discusses this possible explanation for the trends is Blundell and Preston (1995). According to this study, 'the difference between the growth rates of income and consumption inequality is indicative of changes in short-term income risk'. Why should the difference between trends in income and consumption inequality show that there has been increased income risk?

If incomes are more erratic and are subject to more uncertainty than they used to be, then this will probably result in a growth in income inequality. For at any point in time, there are likely to be a number of people whose incomes are unusually high and others whose incomes are unusually low, and, as a result, income differences will be increased. But if people are able to smooth their consumption as their incomes fluctuate, as economic theories such as the permanent income hypothesis[2] suggest (see Chapter 1), then increased income volatility will not result in higher expenditure inequality.

If people do base their consumption decisions on their expected longer-run, or so-called 'permanent', income levels, then so long as the additional income volatility faced by individuals occurs around the same expected *lifetime* levels of income, consumption inequality will not rise. Growing income inequality can be caused both by short-term income fluctuations and by increased lifetime differences, but because of consumption-smoothing behaviour, growing consumption inequality reflects just the changes in lifetime differences.

We have seen in this chapter that both income and spending inequality grew over the 1980s, but that income inequality grew faster. This provides some evidence that there has been an increase in permanent income differences and also a growth in short-term income volatility over and above this. There is some further evidence that can be drawn from our own data to support this hypothesis.

Over the 1980s, there emerged an increased number of relatively high spenders in the lowest income groups, creating the 'tick shape' in Figs. 4.7 and 4.8. Typical spending in the lowest income decile at the end of the period was about 20 per cent higher than typical spending

---

[2] See Friedman (1957).

in the lowest income decile group ten years previously, even though the real level of income in this group was no higher.

Increased short-term income fluctuations could explain this. If the lowest income groups now contain more people who are only at the bottom of the income scale *temporarily*, because of short-term drops in their incomes, then this could explain why their spending is relatively higher. The lowest income groups now contain many more non-pensioners than previously, and it is non-pensioners who are more likely to be reliant on uncertain or irregular streams of income, or who may be at risk of sharp drops in income if made unemployed.

Looking at the bottom income decile group, we see that it is the non-pensioners in this group who have relatively high spending. Figure 4.10 shows how pensioners and non-pensioners in the bottom income decile group are spread across the distribution of expenditure. More than half of all pensioners in this poorest income decile group are to be found also in the lowest spending decile, whereas just 30 per cent of non-pensioners are to be found here. There are a considerable

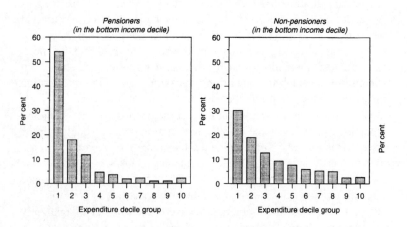

Fig. 4.10. Pensioners and non-pensioners in the bottom income decile group

number of low-income non-pensioners at the top of the expenditure distribution, but very few pensioners.

The low-income self-employed are especially likely to have relatively high spending. Figure 4.11 shows how the various economic status groups in the bottom income decile group are ranked by their spending (full-timers are not shown here because they make up only a very small part of this group). There are more low-income self-employed in the sixth spending decile group than any other, and almost as many as this in the seventh. They are also a group whose incomes can be expected to be particularly erratic or unpredictable,

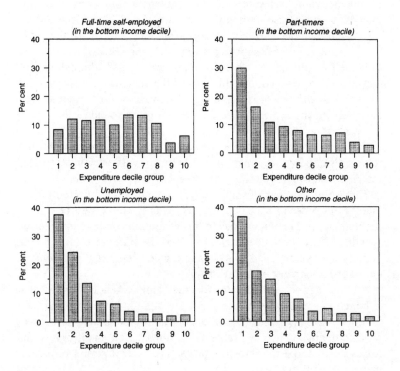

Fig. 4.11. The spending of different groups by economic status in the bottom income decile group, 1991–93

and so if they appear in the bottom income decile group, this may well be because their incomes are only temporarily low. There are also a considerable proportion of part-timers who have relatively high spending compared with their incomes.

Surprisingly, there are not many of the low-income unemployed whose spending is high. One might expect there to be a considerable number of the unemployed whose spells in unemployment are short, who would maintain relatively higher spending throughout their spell of unemployment.

Other Issues

If income volatility were the only explanation for why there are many low-income self-employed in particular with high spending, then we might also expect to see a similar number of the high-income self-employed showing lower spending. But whilst the full-time self-employed on low incomes are spread throughout the distribution of expenditure, those on high incomes are much more likely to also have high spending. The vast majority of the self-employed in the top tenth of incomes are to be found in the top 70 per cent of spenders.

For the self-employed, there is another explanation that springs readily to mind of why there might be many with very low reported incomes: inaccurate reporting of income.

Some self-employed people might intentionally misreport their incomes in order to avoid paying tax. This is unlikely to be a large problem in the data that we use in this chapter to compare income and spending. This is because surveys such as the FES are voluntary, and entirely anonymous, and so systematic under-reporting of income is not likely to be very widespread.

An alternative explanation for the same phenomenon is that although 'normal weekly earnings' are straightforward to measure for those in employment, for the self-employed, finding a measure on a comparable basis is problematic. Information about profits and losses needs to be interpreted, and the convenient period of assessment may well not be a week.[3]

---

[3] Some of the issues involved in measuring the incomes of the self-employed on a comparable basis to the incomes of the employed are considered in Boden and Corden (1994).

Although income from self-employment does account for a growing part of total household income, and so doubts about the comparability of this income to other sources become more relevant, the same trends in income and spending shown in this chapter are apparent when the self-employed are excluded entirely from the population. So although this potential problem should be borne in mind, it should not be given too much weight.

The fact that there are many whose reported incomes are much lower than their spending need not just provide evidence that people are smoothing their consumption between periods of high and low income. It may also be evidence that some families are continually spending beyond their means, and so running themselves into situations of accumulated debt.

There is little evidence from the FES to show that the use of credit expenditure has grown disproportionately amongst the lowest income groups, although it should be noted that the use of credit expenditure is just one aspect of the extent to which families are in debt.[4]

CONCLUSIONS

Some important features of the income and expenditure distributions that have been pointed out in this chapter are summarised below.

1. Inequality of household expenditure grew over the late 1970s and 1980s, but income inequality grew faster.
2. Living standards amongst the 'poorest' tenth of the population rose between 1981–83 and 1991–93 (by 13 per cent) if spending is used to measure living standards, but were roughly constant in real terms if income is used; at the top of the distributions, income grew faster than spending over the same period.
3. Between 1971–73 and 1981–83, it is the income measure of living standards that shows the biggest gains at both the bottom and the top of the distributions.
4. Different groups are most heavily represented amongst the poor depending on whether income or spending is used: pensioners no longer predominate at the bottom of the income distribution, but still remain the largest group amongst the lowest spenders.

[4] This is explored by Ford (1991), who examines the links between low incomes and debt.

5. Alongside the changes in the composition of the lowest income decile group, there has been a growing number of relatively high spenders amongst this group.

6. One important explanation for the different changes in the income and expenditure distributions over the 1980s has been a growth in income volatility or 'income risk' over the period.

5 Accounting for the Trends I: The Sources of Income

INTRODUCTION

Chapter 3 contained an extensive description of trends in UK income inequality over the last three decades. In this chapter, we examine the underlying causes of these trends. For example, we have documented that household income inequality fell between the early and late 1970s before rising substantially over the 1980s. This chapter is the first of three that examine *why* these dramatic changes have taken place. Where do people get their incomes from? How have the different sources of income changed over time, and how has this contributed to the trends in the income distribution that we have seen? These are the issues addressed in this chapter.

So far, we have looked at the distribution of all income, treating income as a single lump. But, of course, people's incomes come from a range of sources such as earnings, social security, and investments, and these are distributed differently throughout the population. The changing sources of income can tell us a good deal about what has been happening to the distribution of income as a whole.

The first sections of this chapter examine the trends in the different components of income and use a measure of inequality that is 'decomposable' in order to assess how changes in different income components have affected overall income inequality. One clear picture to emerge is that earnings make up the largest, albeit a declining, share of total household income.

Because of its importance in total income, inequality in earnings between different households and individuals plays a crucial role in determining the shape of the income distribution. The final section of the chapter therefore focuses on the distribution of earnings, examining the changing nature of the labour market and the working patterns of the population over the last three decades.

THE COMPONENTS OF INCOME

The Current Distribution

For the purposes of this chapter, total income is broken down into six main components:

(1) earnings;
(2) social security benefits;
(3) self-employment income;
(4) private pensions;
(5) investment income; and
(6) other income.

Overall, the most important component of income is earnings, which make up 60 per cent of total net income (in 1992–93), with self-employment income making up a further 9 per cent. Social security accounts for another 18 per cent, investments and private pensions accounting for just over 5 per cent apiece. This composition of incomes is, however, very different at different parts of the income distribution, as shown in Fig. 5.1.

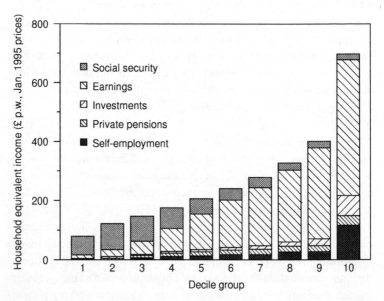

Fig. 5.1. Income composition by decile

For the poorest deciles, earnings play only a very small part, making up just 16 per cent of the incomes of the poorest tenth. This is a reflection of the fact that people in work tend to have enough income to move themselves out of the poorest groups. In the second decile, earnings still provide just under a fifth of total income. They provide somewhat under a half of income in the fourth decile and nearly 60 per cent in the fifth, reaching a maximum of 78 per cent in the ninth. This falls back to two-thirds of total income in the richest decile.

The fall right at the top reflects unusually large receipts of investment income (10 per cent of their total income) and self-employment income (17 per cent) in the top decile. This should not distract from the fact that mean levels of equivalent net earnings are still much higher in the top decile than elsewhere (£463 per week as against £311 in the ninth decile). The unusually high share of self-employment income is associated with the large number of self-employed to be found there—a fact discussed in Chapter 2. The high level of investment income reflects the very considerable concentration of wealth, and therefore income from wealth. The ninth decile has only 6 per cent of its income deriving from investments, other deciles 4 per cent or less.

The gap left by lack of earned income at the bottom of the distribution is filled by social security benefits. The state, through social security benefits, provides 80 per cent of the incomes of the poorest decile, three-quarters of those of the second decile, and well over half (59 per cent) of income in the third decile. It even provides a quarter of the incomes of the fifth decile and 13 per cent of the seventh. But at the top of the distribution, its contribution to the total is negligible, at less than 3 per cent—though this still means an average (equivalent) payment of £18 per week. Much of this will be in the form of Child Benefit and state pensions to the rich pensioners in the top decile. Some benefits are paid to rich households because they contain within them both high- and low-income families, treated as separate units for the assessment of means-tested benefits. For example, an unemployed (grown-up) son or daughter who lives with parents who are earning may qualify for Income Support.

There are many ways of thinking about the inequality of each income component and the extent to which each component adds to or subtracts from overall measured inequality. In the next section, we look at changes in the contribution of different income sources to one

measure of income inequality, half the squared coefficient of variation. Another simple way of looking at the same issue is to consider
the proportion of each income component going to each income group
or, equivalently, average receipt within each group.

Such figures are given in Table 5.1 for each quintile. They show
that the majority of social security income is paid, as one would
expect, to those at or near the bottom of the income distribution,
though more goes to those in the second quintile than to those in
the bottom quintile. This apparently strange finding is largely a reflection of the fact that many individuals are taken out of the bottom
quintile by their social security payments. Those with large amounts
of social security income have enough money to take them out of the
poorest group. This might result from, for example, large amounts of
Housing Benefit or possibly high payments of Invalidity Benefit with
earnings-related additions. As much as 8 per cent of social security
goes to households in the top quintile.

Social security is, though, the odd one out. In all other cases, as
one would expect, richer quintiles take a greater share of the total
income than do lower quintiles. This is most evident in the case of
self-employment and investment incomes, around 60 per cent of
which are received by the richest 20 per cent of households. These
income sources are likely to be particularly disequalising in their
effect. Earnings are not so highly concentrated as these two sources
because they form the main source of income of households in the
middle of the distribution. Nevertheless, very nearly a half of all
earned income is received by the top quintile. Private pensions are
more equally distributed, though the richest group still receives over
a third of the total. Those with high levels of private pension clearly
have enough to get them into the top 20 per cent of the income
distribution, and there are enough of them there to have a high share

Table 5.1. Shares of each income component received by each quintile

	Quintile 1	Quintile 2	Quintile 3	Quintile 4	Quintile 5	
Self-employment	2	8	12	18	58	100
Private pensions	4	13	19	25	38	100
Investment income	3	7	11	17	61	100
Earnings	2	8	17	26	47	100
Social security	29	33	18	12	8	100

of total private pension receipts. In all cases other than social
security, and to a small extent private pensions, the bottom two
quintiles receive only a tiny proportion of the total income. The
reasons for the inequality of the overall income distribution become
increasingly clear.

Changes over Time

Throughout the last three decades, *earnings* have made up the most
important source of total UK household income, but their role has
been a declining one. Figure 5.2 shows how the shares of total household income have altered. Most striking has been the declining contribution of earnings to household income. In the early 1960s, over
three-quarters of all income came directly from earnings. This had
fallen to only about 60 per cent by the start of the 1990s. This reflects
the trend towards lower levels of employment; the greater part of this
fall occurred between 1979 and 1984, during which period the share

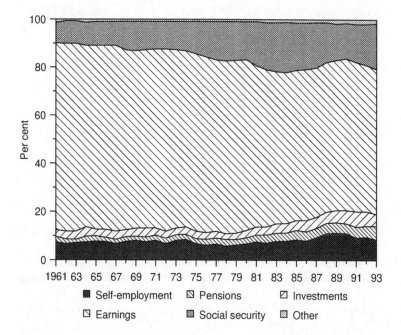

Fig. 5.2. Population income shares

of earnings fell from 71 per cent to 62 per cent, just as the rate of unemployment reached unprecedented levels. But the gradual trend downwards had certainly become apparent before that.

A small part of the fall in income from earnings has been reflected in a rise in income from *self-employment*, which accounted for 11 per cent of income at the end of the 1980s as against 7 or 8 per cent during the 1960s. Nevertheless, the self-employment income share has not risen as fast as the level of self-employment, again reflecting the very low levels of profits made by many of the newest of the self-employed. Since 1989, the share of self-employment income has actually fallen back sharply from around 11 per cent to just over 8 per cent in 1992 and 1993. Both levels of self-employment and levels of profits appear to have suffered in the recession of the early 1990s.

The second biggest source of income throughout the period has been *social security*, and its share has risen sharply from about 10 per cent at the start of the period to 16 per cent in the late 1970s and 20 per cent by 1985. It had fallen back again to 15 per cent of the total by the end of the 1980s, but a very sharp rise at the end saw social security again account for a fifth of all income in 1992 and 1993. For all these detailed changes, the general rise in social security levels over the period is clear. The main reasons for this upward trend have been:

1. *Rising numbers entitled to benefits.* Almost all of the main groups covered by the benefit system have risen greatly in number over the last three decades. These include the unemployed, pensioners, the long-term sick and disabled, and lone parents. The rise in the share of social security in the early 1980s was clearly related to the rise in the numbers of unemployed. The continuing increase in the numbers not in work played a large part in stopping its share from falling back to its original level again as recorded unemployment levels fell.

2. *Rising real housing costs.* Included in this definition of income are benefits that go directly to pay the housing costs of low-income households. In part, therefore, this trend is somewhat artificial since a move from 'bricks-and-mortar subsidy' (for example, low council rents) to a person-based subsidy (for example, Housing Benefit) will raise measured social security income without producing any net increase in living standards. Real spending on Housing Benefit rose very quickly, especially in the early 1990s, not so

much because more people were becoming entitled to the benefit but because rising rent levels in both private and public sectors forced up the benefits of those already receiving Housing Benefit.

3. *The introduction of new benefits.* At the start of the period, the main items included under social security were the National Insurance benefits, National Assistance (the forerunner of Income Support) and Family Allowance. During the 1960s, the various discretionary local systems of Rent Rebates were gradually extended and unified, and a national system of Rates Rebates was introduced in the early 1970s. The early 1970s also saw the introduction of Family Income Supplement (FIS), a completely new benefit for low-waged families with children, and of Invalidity Benefit (at a higher rate than the existing Sickness Benefit). Other benefits for the long-term sick and disabled, such as Attendance Allowance, were added during the late 1970s.

Figure 5.2 also reflects the move in 1977–78 from Child Tax Allowances to Child Benefit, which produces a rise in measured social security income and a fall in after-tax earnings.

4. *Changes in benefit levels.* Over the course of three decades, there are naturally many changes in the value of social security benefits, but two major factors should be noted. The first is the growth of state earnings-related pensions, particularly under SERPS which was introduced in the late 1970s. Pension schemes inevitably take a long time to reach maturity, but the 1980s have brought growing numbers of pensioners receiving significant amounts of earnings-related pension to supplement their basic state pension.

The other important factor is government policy on benefit indexation. Figure 5.2 shows benefits as a share of total household income. Since in almost all years average household income rises faster than prices, benefits would also have to rise faster than prices to maintain their share of total income, other things being equal. In fact, since 1980, the main social security benefits (such as the National Insurance Retirement Pension and Child Benefit) have been increased only in line with prices. As a result, when unemployment is more or less stable (such as between 1983 and 1986), the share of income coming from social security will actually fall. In the period prior to 1980, the state pension tended to rise with the higher of prices and earnings. During the 1960s, the uprating policies tended to be relatively erratic, but on the whole, benefits would rise faster than prices. The change in the indexation policies during the 1980s will have had some effect on slowing down the rise in the importance of social security.

The remaining incomes are *private pensions* and *investment income*. They are not directly earned, though the former in particular can be considered as deferred earnings. Their importance has risen especially since the early 1980s. The increase in private pension income can be largely attributed to the pattern of membership of occupational pension schemes, which reached a peak in the late 1960s before flattening off, improved protection for individuals leaving schemes early, and, of course, higher earnings on which the levels of pension are generally based. We would expect the contribution of private pensions to go on rising.

The contribution of investment income actually declined slightly from the early 1960s to the mid-1970s, but since then has gradually risen to form around 5 per cent of household income in the early 1990s. Two of the main determinants of this trend are interest rate movements and the effects of inflation on the real value of savings. During 1990–91, interest rates reached 15 per cent, and whilst this greatly reduced the after-housing-costs income of net borrowers, it also boosted the income of savers. However, where high interest rates go alongside high inflation, the longer-term effect on investment income is less clear-cut. During much of the 1970s, high nominal interest rates were accompanied by even higher inflation which eroded the real value of savings. As a result, the capital on which interest was being earned was being gradually eroded. Those living on the investment income would have seen this process happening particularly quickly.

The composition of incomes by income range has also changed over time. Income shares over time for the first, fifth, and tenth deciles are shown in Fig. 5.3.

Even in the bottom decile, the proportion of income made up of earnings has fallen from an average of around 17 per cent of the total in the 1960s to about 12 per cent in the late 1980s and early 1990s. Social security income has seen a corresponding increase in its share of income in this poorest group. No clear trend is perceptible in the small share taken up by other income types. All in all, the picture has not changed dramatically over time, with state benefits forming a substantial majority of the income over the whole period.

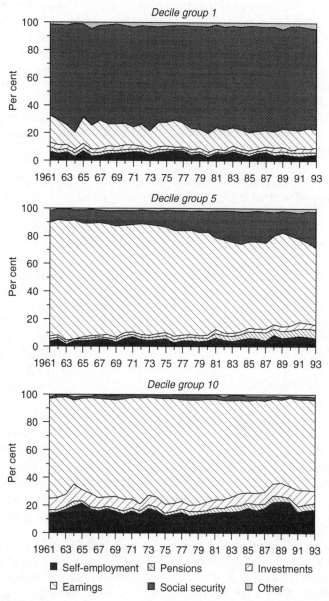

Fig. 5.3. Shares of total household income by source and income range

Patterns in the middle of the distribution are more dramatic, with the share of earnings falling from over 80 per cent of the income of the fifth decile at the beginning of the period to nearer 60 per cent by the early 1990s, while social security's share rose from under 10 per cent at the beginning to over a fifth in the early 1980s before falling back in the boom at the end of the 1980s and rising again in the early 1990s. Both pensions and investment incomes also saw their share in the incomes of this middle decile rise as pensioners moved up into the middle of the income distribution as a result of their higher private incomes. From accounting for around 3 per cent of the income of the fifth decile in the 1960s between them, their combined share had risen to 8 or 9 per cent in more recent years.

The trends, such as they are, at the top of the income pile are rather unlike those elsewhere in the distribution. At no point over the period did social security benefits account for more than 4 per cent of the total income of this group. The share of earnings, following a small rise to a maximum of 75 per cent in the late 1970s, actually fell over the whole period to a minimum of near enough 60 per cent at the end of the 1980s and two-thirds in the most recent years. Their place was taken in part by each of the other income sources. The importance of investments dipped noticeably in the late 1970s, but climbed during the 1980s, while private pensions' share roughly doubled from about 2 per cent of the total in the 1960s to about 4 per cent at the end of the period.

INEQUALITY AND INCOME SOURCES

We have seen in Chapter 3 that household incomes have become increasingly unequally distributed, especially during the 1980s. But household income is merely the sum of the various components of income that we have been considering in this chapter. Some sorts of incomes are more unequally distributed than others, and the changes are different over time.

In order to examine the extent to which the different income components contribute to income inequality, we need to employ a different measure of inequality from the ones that we have been using so far. Following the methodology of Shorrocks (1982a), as applied by Jenkins (1995), we use a measure that is half the squared coefficient of variation. This measure can be readily 'decomposed' or broken down into its constituent parts.

Before we go on to illustrate what the decomposition of the index can show us about changing income inequality, we compare half the squared coefficient of variation with the Gini coefficient (which we have used in previous chapters as a summary measure of the level of inequality). Figure 5.4 shows how moving averages of the two measures compare. The units that the two are measured in are not relevant for this comparison—it is the changes over time that are important—and so Fig. 5.4 rebases both of these to equal 100 in 1961–63. What is immediately obvious from this figure is that the magnitudes of the changes in half the squared coefficient of variation are larger than the magnitudes of the changes in the Gini coefficient. This is partly because this measure is not bounded by 0 and 1 like the Gini, and so proportional changes are likely to be larger. It is also more sensitive to outliers than is the Gini coefficient. Nevertheless, the direction of the changes in the two measures is the same.

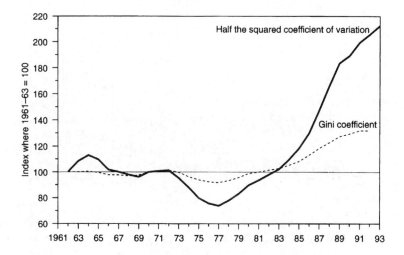

Fig. 5.4. The Gini coefficient and half the squared coefficient of variation compared

Note: Three-year moving averages.

This is important, because when we explain what has happened to half the squared coefficient of variation by decomposing it into its constituent parts, we want this explanation to apply more generally to what has happened to the distribution as a whole. Because the sorts of movements we see in this measure are similar to the movements in other inequality measures, we can be confident that we are explaining a wider phenomenon rather than just the features of one particular inequality measure. However, the differences between the two measures should be borne in mind.

We next look at the part played by each different component of income in explaining these movements in inequality. The contribution of each income source (earnings, pensions, investments, etc.) to total inequality as measured by half the squared coefficient of variation is shown in Fig. 5.5. The sum of each of the component parts gives the total.

The first message of Fig. 5.5 is that of all the components of household income, only social security has an equalising effect. In

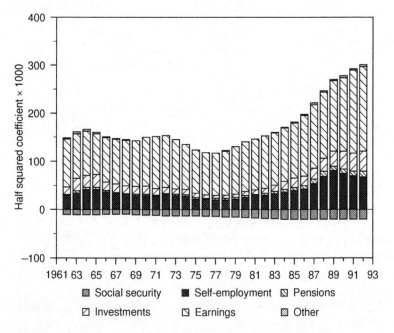

Fig. 5.5. Decomposition of overall income inequality by income source

other words (and hardly surprisingly), if the social security system were to be abolished, the immediate effect would be to make the income distribution even more unequal than it already is! More interestingly, the equalising properties of the benefit system are shown to have increased slightly up to the early 1980s and to have levelled off thereafter.

Amongst the remaining sources of income, all of which contribute to overall inequality, clearly earnings inequality is the most important. The trends in earnings inequality are discussed in more detail in the next section.

But whilst changes in earnings inequality more or less explained the trend in overall inequality for much of the 1960s and 1970s, earnings inequality is far from being the whole story of the 1980s. Although earnings inequality has been rising, the inequality accounted for by other income sources has grown faster, and so the *proportion* of overall inequality that is accounted for by earnings has actually fallen as overall income inequality has grown over the last fifteen years.

In the second half of the 1980s in particular, the inequality of self-employment incomes has greatly increased and has contributed to the rise in overall inequality. The growing differentiation in the incomes of the self-employed has been referred to earlier when explaining the fact that the role of self-employment income had not grown as rapidly as the growth in the numbers of self-employed might suggest. We discuss some possible reasons for this in Chapter 4, when we compare the spending of the self-employed with their incomes. The other major contributor to the growth in overall income inequality has been investment income.

We can interpret the trends from Fig. 5.5 more precisely by understanding the way in which each income source contributes to the overall inequality measure. This will enable us to understand, for example, *why* it is that earnings make up the biggest contribution to overall inequality, why social security has an equalising effect, and what has driven the changes in the 'composition' of inequality over time.

The importance of each income source in contributing to overall inequality is determined by three separate factors, which are outlined below.

1. *Within-source inequality (A)*. The more unequal the distribution of each source of income (amongst the entire population), the higher is its contribution to overall inequality.

   For sources of income that are received by a few people but not received by many others, we would expect the 'within-source' inequality to be high. Income from private pensions are an example of this.

   Similarly, for sources of income that some people receive in large amounts but others only receive as small top-ups to their incomes, we would also expect the within-source inequality measure to be higher than it is for income sources such as earnings which tend to be the sole, or at least a major, source of income for those who receive them. Investment income falls into this category. The majority of the population are in households receiving some investment income, but many of them receive only small amounts, for example in the form of interest on their savings accounts. On the other hand, there are also a small number of wealthy households that receive substantial income from their investments. Measured inequality in investment income is therefore very high.

2. *The share of the source in total income (B)*. The higher the share of each income source in total income, the larger its contribution to overall inequality. For example, in the last section, we saw that earnings take up the largest share of total income. It is partly for this reason that its contribution to overall inequality is high.

3. *The correlation of the source to total income (C)*. If the source is positively correlated to total income, this means that, in general, high-income households receive more income from that source than do low-income households. The higher the correlation, the more that income source contributes to overall inequality.

   Social security is the only source of income that is negatively correlated to total income: receipt is concentrated at the bottom end of the income distribution. This is why social security makes a negative contribution to overall inequality.

We can now explain the trends in inequality illustrated in Fig. 5.5 in the light of changes in each of the different factors set out above. For reference, the overall level of inequality in three selected groups of years is shown in Table 5.2. A detailed breakdown for each of the different income sources is provided in the different parts of Table 5.3. The total contribution of each source, presented in bold in these

Table 5.2. Overall income inequality measured by half the squared coefficient of variation

	1961–63	1976–78	1991–93
Half the squared coefficient of variation × 1000	137	101	280

parts, is in fact the product of four values: the three factors (A), (B), and (C) discussed above, which are also given in the parts of Table 5.3, and the square root of the figure for overall inequality given in Table 5.2 (see Chapter 1 for further explanation of this breakdown).

Earnings

Earnings make up the largest single contribution to overall inequality, accounting in 1991–93 for 177 of the total value for half the squared coefficient of variation of 280 (or 63 per cent) (see Table 5.3(a)). The reason why earnings make up the largest single contribution to overall inequality lies both in the large share of earnings in total income (61 per cent in 1991–93) and in the high correlation of earnings with income. The within-source inequality for earnings is in fact relatively low compared with within-source inequality for the other income sources.

Although the contribution of earnings to the total has risen, the importance of earnings in explaining overall income inequality has declined over time. The *proportional* contribution of earnings to total inequality was at its highest at the end of the 1970s. In 1976–78, earnings accounted for a full 88 out of the level for total inequality of 101 (87 per cent).

Table 5.3(a). Detailed decomposition of inequality: earnings

	1961–63	1976–78	1991–93
Earnings	**100**	**88**	**177**
(A) Within-source inequality (square root × 1000)	501	536	802
(B) Factor share	0.77	0.72	0.61
(C) Correlation with total income	0.69	0.72	0.69

The declining proportional contribution of earnings is explained by the fall in the share of earnings in total income which was discussed in the previous section. This drop, from almost 80 per cent in the 1960s to just over 60 per cent in the 1990s, has meant that, despite rising 'within-source' earnings inequality and a high positive correlation of earnings with total income, the total contribution of earnings to the overall inequality measure has not risen as fast as has the contribution of other income sources.

Social Security

Social security is the only source of income to make a negative contribution to overall inequality. As can be seen in Table 5.3(b), the reason for this negative contribution is that social security benefits are negatively correlated with total income. This means that if social security payments were abolished, income inequality would be higher than it is at the moment.

Social security benefits make up the next biggest share of total income, after earnings, and this is the main factor accounting for the size of the equalising effect that social security has. Within-source inequality is relatively low compared with other income sources, since receipt of benefits tends to be clustered around a few different benefit levels.

As for the changes over time, we saw from Fig. 5.5 that the equalising properties of the social security system rose slightly up to the early 1980s. This rise was driven by the growing share of social security in total income that we discussed above. This growth in the factor share was somewhat offset over the 1980s by falling within-source inequality and by a correlation of benefit income with total income that has become slightly less negative over time. This is

Table 5.3(b). Detailed decomposition of inequality: social security

	1961–63	1976–78	1991–93
Social security	**−11**	**−15**	**−21**
(A) Within-source inequality (square root × 1000)	1010	908	805
(B) Factor share	0.094	0.158	0.175
(C) Correlation with total income	−0.32	−0.34	−0.29

because the basic state pension now goes to many households also receiving income from other sources, such as private occupational and personal pensions.

Self-Employment Income

Self-employment income accounted for 67 out of the total value for inequality of 280 in 1991–93 (24 per cent). Although the share of self-employment income in total income is relatively low, at about 9 per cent in 1991–93, this is offset by a relatively high correlation with total income and a very high measured within-source inequality.

The reason for this high measured inequality in the receipt of self-employment income lies partly in the fact that smaller numbers of people receive any income at all from self-employment, as compared with earnings or social security. Amongst those who do, many receive small supplementary amounts from part-time self-employment possibly in addition to employed work undertaken elsewhere, whilst yet others are making zero profits or reporting losses. At the same time, there are significant numbers who are reporting very large self-employment profits. This means that the measured dispersion in self-employment incomes is very large.

The growing importance of self-employment income in contributing to overall inequality is the result of a rise in all three of the factors that were outlined above: within-source inequality has risen, self-employment income has become more positively correlated with total income, and the factor share has grown.

Table 5.3(c). Detailed decomposition of inequality: self-employment income

	1961–63	1976–78	1991–93
Self-employment income	**29**	**19**	**67**
(A) Within-source inequality (square root × 1000)	3067	3118	3247
(B) Factor share	0.072	0.062	0.090
(C) Correlation with total income	0.35	0.30	0.44

Private Pensions

The contribution to total inequality from private pensions has risen rapidly over the 1980s and early 1990s. In 1991–93, private pensions income accounted for 12 out of the total inequality measure of 280, as compared with only 2 out of 137 in 1961–63. This has arisen because of the growing amount of income from private pensions in the total. The share of private pensions in total income has risen from less than 2 per cent in the 1960s, to 5 per cent by the early 1990s. Pension income has been increasingly positively correlated with total income: additional private pension receipt has tended to go more to households with higher total incomes.

Measured 'within-source' inequality for private pensions has actually fallen over the period. As more people have started receiving income from private pensions, the inequality measured across the population has dropped.

Investment Income

Investment income is another source of income whose contribution to overall inequality has risen dramatically over the 1980s. This is because the number of households receiving large amounts of investment incomes, from property, share income, and other more large-scale investments, has risen. This has meant that investment income has become increasingly positively correlated with total income: it is high-income households in which the receipt of large amounts of investment income is concentrated. The share of investment income in total income has also risen.

Measured within-source inequality in investment income is high compared with many other income sources. This is because there is a

Table 5.3(d). Detailed decomposition of inequality: private pensions

	1961–63	1976–78	1991–93
Private pensions	**2**	**3**	**12**
(A) Within-source inequality (square root × 1000)	3608	3087	2476
(B) Factor share	0.018	0.024	0.050
(C) Correlation with total income	0.09	0.13	0.18

Table 5.3(e). Detailed decomposition of inequality: investment income

	1961–63	1976–78	1991–93
Investment income	**17**	**7**	**42**
(A) Within-source inequality (square root × 1000)	4877	3075	3094
(B) Factor share	0.031	0.026	0.058
(C) Correlation with total income	0.30	0.28	0.44

wide dispersion between the investment incomes of those who receive small amounts of income on their savings accounts and of those who receive substantial income from their large-scale investments. The within-source inequality of investment income fell over the 1960s and early 1970s, but has not changed significantly over the 1980s.

EARNINGS

The rest of this chapter will be concerned with the distribution of earnings. The contribution of earnings to total income, and of earnings inequality to overall income inequality, has declined, particularly over the 1980s. But earnings are still by a large distance the biggest single source of income, and make by far the biggest contribution to overall inequality. There is a large literature, both in the UK and in the US, that examines and explains the changing distribution of individual earnings over time; the explanations put forward to account for the changing distribution of individual earnings will clearly play an important part in explaining the changes of the distribution of household income as a whole.

The way in which earnings are distributed across households, and how this contributes to household income inequality, depend upon a range of different factors, including

(1) how many people in the household work;
(2) the number of hours for which they work;
(3) the hourly wage; and
(4) since income is measured net, the amount of tax that they pay.

The first three of these we will consider in this chapter. The way in which the tax system affects different households is the subject of

Chapter 7. We begin by examining the way in which participation in the work-force has changed.

Participation and Hours

The last thirty years have seen important trends in the working patterns of the population, both in terms of the sorts of people who work and in terms of the number of hours for which they work. In order to examine these issues, this section looks at the changing numbers of people in full- and part-time employment (or self-employment) over time.

The major story is that the participation patterns for women and men are very different: the participation of women in the work-force has increased, whilst the proportion of working-aged men who are in employment has fallen considerably.

A large proportion of women who are in employment work part-time, and so the balance between part-time and full-time work in the economy has changed. Much of the increased female participation has also been from married women whose husbands also work. This has led to an increasing polarisation between households that have two earners and those that have no earners at all.[1]

In this section, we look at the changes in male and female participation, by age, marital status, education group, and region. By participation, we mean the proportion of the working-aged population who are in employment or self-employment. We then look at how the distribution of work between different households has changed, and discuss some studies that have examined the impact of this on household income inequality.

Consider first the overall participation of men and women in the work-force. Figure 5.6 shows the proportion of men and women who are in employment or self-employment. (We consider those who have left school, so are no longer classified as dependent on their parents, and under the state pension age.) Note that at this stage we do not distinguish in any way between different reasons for being out of work: amongst those not participating will be people who are unemployed and seeking work; others will be out of work but not seeking work; yet others will be long-term sick or disabled, early retired,

---

[1] See Gregg and Wadsworth (1996).

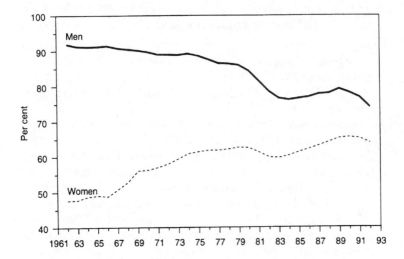

Fig. 5.6. Percentage of working-age population in employment or self-employment

students in further education, on government training schemes, etc. We return to the changing reasons for non-participation below.

A clear pattern emerges from Fig. 5.6. Over the whole period, female participation has risen whilst male participation has fallen. Most of the rise in female participation in fact took place during the 1960s and 1970s, falling off slightly over the recession during the late 1970s and early 1980s, and then rising again over the mid- to late 1980s. The recession over the early part of the 1990s has seen some levelling off in the rate of female participation.

By contrast, male participation was a fairly constant proportion of the male population over the 1960s and 1970s, but fell sharply over the late 1970s and early 1980s. It continued to fall throughout the 1980s and early 1990s. By the end of the period, the male and female participation rates were much closer together than at the start of the period. In the early 1990s, about 74 per cent of working-age men were in employment or self-employment, and about 64 per cent of women. By contrast, in the early 1960s, as many as 92 per cent of men were participating in the work-force, compared with 48 per cent of women.

The pattern of work and non-work has not been the same amongst all groups of men and women. Some important differences in the trends in the participation of different groups are discussed below.

1. *Age*. Across most age-groups in the population, the general pattern of increasing female participation over the 1970s and falling male participation over the 1980s is borne out. However, the pattern amongst the youngest and the oldest groups deserves particular attention.

Amongst both men and women, participation rates for *younger groups* of the population, in particular for 16- to 19-year-olds, have fallen off over the last three decades. Although the fall has been fairly steady over time, the sharpest drop came during the recession of the early 1980s, and a somewhat smaller one over the recession of the early 1990s. Although there is higher claimant unemployment amongst the young, two of the reasons for the falling participation of this younger group have been an increase in the take-up of higher and further education, and the introduction of government training schemes such as the YTS.

Amongst men, participation is the lowest, and has fallen off the most dramatically, in the age-group that is *just below retirement age* (60- to 64-year-olds). Participation amongst those aged 55–59 has also fallen back considerably over the 1980s. In the recession of the early 1980s in particular, the number of men in early retirement rose sharply. Amongst women just below the pension age, participation rose steadily over the 1970s, but has fallen off during the 1980s, although not to the same extent as amongst older men. Some studies have examined the extent to which the phenomenon of early retirement over the 1980s can be interpreted as a growth in 'disguised unemployment'.[2]

2. *Region*. There is also some regional variation in the changes in participation rates. At the start of the 1960s, male participation rates were at similar levels across the different regions of Great Britain.[3] From about the mid-1970s onwards, participation rates in the North, in Scotland, and in Wales dropped below those in the other regions, and fell very steeply indeed over the recession of the

----

[2] For example, see Disney, Meghir, and Whitehouse (1994).

[3] We do not consider participation in Northern Ireland in this section because there is a possibility of sampling error in using the Northern Irish Family Expenditure Survey data in isolation from the other regions of the UK. This is because up until the early 1990s, the response rate to the Northern Irish FES was falling over time.

early 1980s. The male participation rate in the Midlands also fell back steeply at this time.

Over the 1980s, participation in Greater London and the South was higher than elsewhere in the country. The recession in the early 1990s saw a particularly sharp drop in male employment and self-employment in the Greater London area. By the end of the latest recession, it was only the South that had significantly higher participation rates amongst men than those in the rest of the country.

Although there is some regional variation in the participation rates of women, the changes over time have been fairly similar across the different regions. Female participation has been higher in Greater London and the South than elsewhere in the country, and lower in Wales throughout the period under examination. The last recession saw a dip downwards in the participation of women (as well as men, as discussed above) in Greater London.

3. *Family type*. Falling male participation has occurred across all groups of men when classified both according to their marital status and according to whether or not they have children. But the drop in participation is particularly marked amongst those men who are married whose partners do not work. Male participation is highest, and the drop in participation less pronounced, amongst men who are married and whose spouses also work, and also amongst single men without children. Lone fathers make up a very small proportion of the population, but amongst this group, participation is very low.

Again amongst women, those who have working spouses have higher and faster-growing participation rates than those who are married with a husband out of work, so that the number of married couples in which the sole breadwinner is the female partner is very small. The participation of single women with no children has actually fallen off since the late 1970s, as has lone-parent participation.

4. *Education*. The biggest drop in male participation has been amongst those who left school at the minimum school-leaving age, whilst the biggest increases in female participation have been from more-educated rather than less-educated women. One explanation that has been put forward for this is that the job opportunities for those with lower skills and education have fallen, particularly over the 1980s, resulting in higher unemployment and

also lower wages amongst the unskilled. We return to this issue when considering the changes that have taken place in the distribution of wages.

There are also important differences in the number of *hours* that men and women who are in employment or self-employment tend to work. Of those men who are in work, very few indeed work part-time. Almost all work in the region of 40 hours per week, and this has changed little over time.

By contrast, the distribution of women's working hours is very dispersed. Amongst working women, a large proportion work part-time, and this proportion has risen as more women have entered the labour force. Part-time workers tend to work a wider range of different hours than full-timers, although there is a noticeable cluster at about 20 hours per week.

How do these patterns of participation affect the way in which work is spread between households? One clear pattern to emerge has been a growing polarisation between households that contain two earners and those in which there are none, or those that are 'work-rich' and those that are 'work-poor'.[4] As was charted in Chapter 3, the 'traditional' single-earner couple has been in decline. Over the 1960s, more than one-third of the whole population lived in a family in which there was one breadwinner working full-time and one unoccupied spouse. By the end of the 1970s, this proportion had shrunk to less than a quarter, and by the early 1990s to around 15 per cent.

Part of this decline is accounted for in a growth in the number of two-earner couples. Increased female participation in the labour force has come predominantly from married women whose husbands also work. At the same time, those whose husbands do not work are likely also to be out of work themselves. The reverse side of the same coin is that the decline in male participation has come largely from men whose wives do not work either.

One possible reason for this is that people often marry those with similar skills and educational background to themselves. If this is the case, then when one partner is unable to find work, it is quite likely that the other partner will not be able to find work either. The fall in the demand for unskilled workers (which is discussed more fully below), and the increased demand for skilled workers, will have affected both partners in many couples in the same way: either both

---

[4] These phrases were coined in Gregg and Wadsworth (1996).

partners have been able to find work or neither of them have. In the case where neither partner in the same household has work, this is then a result of 'a coincidence of members experiencing common adverse trends in the labour market' (Gregg and Wadsworth, 1996). This explanation applies equally to couples in which both partners have been made unemployed and to the many cases where the wives of older, unskilled men have never joined the work-force at all.

Another reason could be the way in which the benefit system works. Means-testing for benefits such as Income Support is based on the income of the entire family. If one spouse goes out to work, this may mean that the family loses its entitlement to benefits, clearly reducing the incentive for either partner to take up work, particularly if it is low-paid.

Figure 5.7 shows the proportion of the non-pensioner population living in families (defined as head, spouse, and any dependent children) containing either zero earners (or self-employed), one earner, or two earners. It shows that there has been a growth in the number of two-earner families, particularly over the 1970s when the increase in female participation was at its fastest; it also highlights the growth in

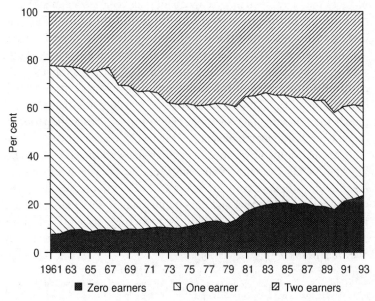

Fig. 5.7. Percentage of population in families with zero, one, or two earners

the number of families with no earners, particularly since the late 1970s and early 1980s when unemployment was rising. The result has been a squeeze in the number of people living in families with just one breadwinner.

For those who are single, the highest number of earners there could be in the family unit is one; the trends in Fig. 5.7 will be partly a reflection of changing family structures, as well as partly a reflection of the patterns of participation in the work-force. For this reason, Fig. 5.8 breaks down the non-pensioner population into married and single people. Figure 5.8(a) shows the number of workers in married couples, whilst Fig. 5.8(b) shows the split between single adults in employment and those not working.

There have been some studies that have examined the extent to which this division between work-rich families with two earners and work-poor families with no earners at all has contributed to the growth in household income inequality. For example, Borooah and McKee (1995) decompose the squared coefficient of variation into various income components, treating husbands' and wives' earnings as separate income sources. In doing this, they are able to look at how much of the increase in income inequality between 1979 and 1993 has been a result of changes in wives' earnings over the period. The main finding of this research is that wives' earnings have *not* played a major part in the growth in household income inequality that we have seen.[5]

If we look again at the changing patterns of participation and non-participation over time described above, the reason for this finding becomes apparent. The major growth in female participation in the labour force in fact came over the 1970s, *prior* to the main period of rising household income inequality. Figure 5.8 makes it clear that the main reason for the growing division between working and non-working households over the 1980s has been the growth in zero-earner households as opposed to a growth in two-earner households. Clearly, the rise in unemployment and non-employment, rather than increased female participation in households with working husbands, will have a larger part to play in explaining the growth in income inequality over the 1980s.

[5] Harkness, Machin, and Waldfogel (1996) find (over a similar time period, 1979–81 to 1989–91) that amongst married couples and in the population as a whole, women's earnings have had a somewhat equalising impact on the distribution of income.

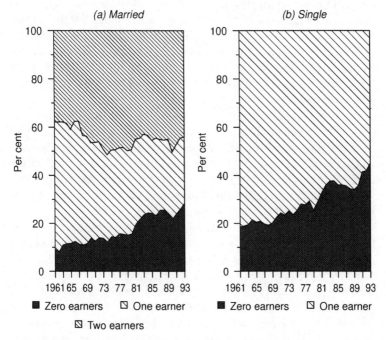

Fig. 5.8. Percentage of population in families with zero, one, or two earners: married and single

The Distribution of Wages

The pattern of wage dispersion over the last three decades shows considerable similarities to the trends in overall household income inequality; together with the decompositions shown earlier in this chapter, this suggests that the distribution of hourly earnings plays an important role in shaping the income distribution. The wages of men and women show some distinct trends over time, and so we examine them separately.

Looking first at the wages of *men*, we consider evidence from Gosling, Machin, and Meghir (1994), who have used information in the Family Expenditure Survey about the hourly earnings of men aged 23 to 59 between 1966 and 1992.[6] They have compared the 90/10

---

[6] The analysis begins in 1966 rather than 1961 because detailed information about hours of work is unavailable prior to 1966.

ratio for male hourly earnings with the 90/10 ratio for the exact definition of net household equivalent income that we use in this book. These are shown in Fig. 5.9. The changes over time are broadly similar. In particular, it can be seen that wage dispersion as measured by the 90/10 ratio fell over the earlier part of the 1970s, and has grown since 1976; income inequality was falling over the 1970s until 1977, and has grown thereafter.

There are some important differences in the trends. First, the growth in income inequality slowed somewhat over the early part of the 1980s whereas wage inequality continued to grow. Second, inequality in our household income measure has grown much faster over the later part of the 1980s than has wage dispersion. This provides further evidence that changes in earnings do not provide the whole story about income inequality over the 1980s.

Although Fig. 5.9 shows a falling 90/10 ratio for male wages in the late 1960s to mid-1970s, wage relativities were in fact remarkably constant over this time. Figure 5.10(a), also from Gosling, Machin, and Meghir (1994), shows how real hourly wages for the tenth percentile, the fiftieth percentile, and the ninetieth percentile of male earners have changed between 1966 and 1993. The wages of each are presented as an index, with 1966=100; this gives a clear picture of differential wage *growth* at different parts of the distribution. The authors have pointed out that four different stages emerge over the period.

The first, between 1966 and 1972, was one where wages across the distribution were growing at a similar rate. The second stage, 1972–75, was one in which wages at the bottom of the distribution grew faster than those at the middle, which in turn grew faster than wages at the top, so that wage inequality between male workers was falling somewhat. The third period, 1975–77, saw falling wages across the distribution, with the top being affected the most. From 1978 to 1993, wage differentials clearly grew. Whilst the tenth percentile wage barely changed, the median wage increased by about 30 per cent and the ninetieth percentile by more than 50 per cent.

Women's wages are typically lower than men's, although the gap between women's and men's hourly wages has narrowed over time. In the late 1960s, women's hourly earnings were typically[7] only about 55 per cent of men's. The early to mid-1970s saw a considerable

---

[7] The ratio of men's median to women's median earnings.

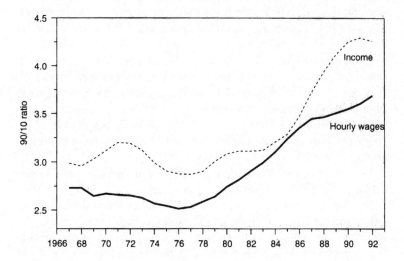

Fig. 5.9. The male wage distribution and the income distribution

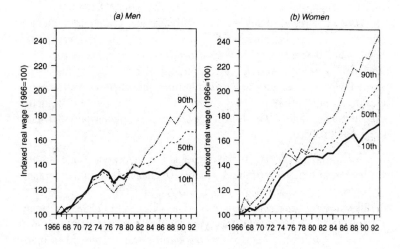

Fig. 5.10. Tenth percentile, median, and ninetieth percentile wages

narrowing of this gap; by the end of the 1970s, the ratio was about 65 per cent. Over the 1980s, the gap continued to decline, but at a slower rate than over the decade before: by 1993, this ratio was at about 70 per cent.

It has been pointed out by Harkness, Machin, and Waldfogel (1996) that most of the narrowing of the gap has come from improvements in the wages of women who work full-time. Using data from the General Household Survey, they show that the ratio of women's to men's wages for full-timers aged 24 to 55 rose from 70 per cent in 1979 to 75 per cent in 1989, whereas for part-timers this ratio rose only slightly, from 55 per cent to 57 per cent.

The trend in the distribution of women's wages has been rather different from that of men's wages and can be seen in Fig. 5.10(b). First, there has been a changing wage structure throughout: this is unlike the male wage distribution where relativities did not change in the early part of the period. Over the period 1976–78, wages of the ninetieth percentile and the median fell, but the tenth percentile wage continued to grow. Most strikingly, over the 1980s, the tenth percentile of women's wages continued to grow, whilst that of men's wages stagnated. Between 1978 and 1993, the tenth percentile female wage grew by about 23 per cent, compared with almost 38 per cent growth in the median and 61 per cent growth in the ninetieth percentile.

There have been many studies, both in the UK and in the US where a similar phenomenon of growing male wage dispersion has been seen, that have pointed to a drop in the *relative demand for unskilled workers*, and a corresponding rise in the demand for skilled workers, as the key to explaining the changes in the male earnings distribution that have taken place. What forces might be driving such a shift? The following are some explanations that have been put forward:

1. *Technological change* in favour of skilled workers: for example, the computerisation of many workplaces has increased the demand for skilled workers at the expense of the unskilled.

2. *Structural shift in employment from manufacturing to services* may have led to a change in the types of workers that firms require. Only about 18 per cent of the work-force was employed in manufacturing in 1991, compared with 37 per cent in 1961.

3. *Increased international trade*, especially from developing countries: as trade barriers have come down, unskilled workers in the

UK have been placed in competition with the large supplies of relatively cheap, unskilled labour available in developing countries; this competition works either directly, if international firms are able to choose whether to employ workers in the UK or abroad, or indirectly, through competition in goods produced by unskilled workers.

There is some debate about which of these factors is the most important in driving the changes. Some have pointed to the fact that the changes in demand for skilled and unskilled workers have not just taken place between manufacturing and services but have also been a phenomenon within manufacturing. This would indicate that the shift in production away from more traditional manufacturing industries and towards services cannot account completely for the changing demand for workers of different skills. Similarly, increased competition from abroad as a result of more open trade can only explain part of the change that has occurred, since it is not just those industries whose products are subject to international competition that have cut back their demand for unskilled workers.[8]

Why would such a shift in demand away from unskilled workers result in the widening wage inequality over the late 1970s and early 1980s? It is because skills confer a premium to people's wages, and this premium can change over time depending on the demand for and the supply of these skills. Taking into account other factors such as differences in age and work experience, many studies have found that the wages that unskilled workers can command have fallen, particularly for those younger workers who are recent entrants into the labour force, whilst the premium to wages that skilled workers can command has risen.

Figure 5.11 shows demand and supply curves for unskilled labour. We know that the real wage going to unskilled labour has fallen. This cannot be explained by an outward shift in supply of unskilled workers, since the number of people continuing in education beyond the ages of 16 and 18 in the UK has risen, if anything actually reducing the supply of less skilled workers. In a competitive market, the explanation for falling wages for unskilled workers must therefore lie in the fact that demand for such workers has fallen. This is shown by the inward shift of the demand curve in Fig. 5.11 from D1 to D2.

---

[8] See Burtless (1995), Schmitt (1995), and Juhn, Murphy, and Pierce (1993) for a good discussion of these issues.

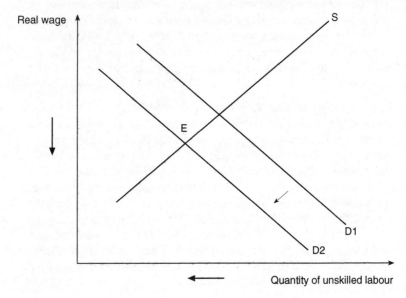

Fig. 5.11. Demand and supply curves for unskilled labour

Note that such a shift can also explain why employment levels amongst those with lower education have dropped. The new equilibrium, at E, is now both at a lower wage and at a lower level of employment.

On the other hand, this cannot be the whole explanation for why earnings inequality has grown. Whilst it has been found that the difference in wages *between* workers of different skills has grown over the 1980s, it has also been found that inequality *within* groups of people with the same skills has also grown. In particular, wage inequality has grown amongst the low-skilled. For example, Juhn, Murphy, and Pierce (1993) found that only about one-third of the rise in male wage inequality in the US between 1963 and 1989 can be accounted for by growing disparities between identifiable groups (such as those with university degrees). Similarly, studies of wage inequality in the UK such as Schmitt (1995) and Gosling, Machin, and Meghir (1994) have also found considerable growth in inequality within groups of similar sorts of workers.

Shifts in the demand for workers with different levels of skill cannot explain this portion of the rise in inequality.[9] At the same time, there have been many 'supply-side' or institutional factors that are also likely to have had an effect on the distribution of earnings between different people. Some of the most important of these are outlined below.

1. *Incomes policies*. Over the 1960s and 1970s, the government operated a series of policies that were designed to control inflation by restricting wage growth. The imposition of these, and their subsequent removal, or, more often, collapse, can be expected to have had an impact on wage differentials. This is particularly the case for those policies that contained a flat-rate element, since the same cash increase represents a higher percentage increase for the less well paid than for the more highly paid.

In the case where prices are rising, a particular cash increase might even result in real income gains for those at the bottom of the distribution by increasing their incomes by more than inflation, while at the same time resulting in real income falls for those at the top of the distribution. A wage standstill as opposed to a flat-rate cash rise should have an equal effect on the real wage for all earners: the real wage should fall exactly in proportion to the rate of inflation, and rather than eroding income differentials, it should hold them constant between earners.

Box 5.1 sets out the major incomes policies that were put into effect over the 1960s. Those with some sort of flat-rate element are highlighted in bold. Although we do not know the exact contribution that these policies made towards the changes that took place in the wage structure, at least some of the wage compression over the mid-1970s is considered to be a direct result of the incomes policies in place at the time.[10]

2. *Equal Pay Act*. This was passed in 1970 and implemented in 1975. Employers were thus given a five-year adjustment period to comply with the equal pay measures specified. Over this adjustment period, as we have seen, the gap between male and female hourly wages fell considerably.

3. *Wages Councils*. Although the main element of government statutory control over pay ended with the abandonment of incomes

---

[9] Although it could be the result of changing returns to skills that are unobservable in the data.

[10] For example, see Ashenfelter and Layard (1983) and Schmitt (1995).

Box 5.1. Incomes policies, 1961–78

Those policies involving some flat-rate element are highlighted in bold.

1960s:

July 1961   **Non-statutory 'pay pause'.**

Oct. 1962   Voluntary 'norm' of 3–3.5 per cent for annual wage increases set by National Incomes Commission (NIC).

March 1965   Voluntary 'norm' of 3–3.5 per cent for annual wage increases set by National Board for Prices and Incomes (NBPI).

July 1966   **Statutory pay freeze imposed for 6 months.**

Jan. 1967   'Severe restraint' with a 'nil norm' and pay rises only if linked to productivity improvements or justified by criteria for exceptional treatment.

1968–69   3.5 per cent ceiling, replaced at end of 1969 by 2.5–4.5 per cent.

1972–74:

Nov. 1972   **Standstill on wages, dividends, rents, and prices.**

March 1973   **Annual increases limited to £1 p.w. per employee + 4 per cent of the employer's pay bill per employee in the preceding year up to a maximum of £250 p.a. per individual;** this rule defined the total sum available for a group of employees; how it was divided up between them was determined by a process known as 'kitty bargaining'.

Nov. 1973   **Ceiling of 7 per cent of the group pay bill (excluding overtime) or £2.25 per person p.w., to a limit of £350 p.a. per individual** and with provision for productivity agreements and indexation: wage increases triggered automatically if the price level rose above a designated threshold.

1975–77: The 'Social Contract' era

July 1975   **Pay increases restricted to £6 p.w., with no increase at all for those above £8500 p.a.**

July 1976   **Pay increases restricted to 5 per cent, with a lower limit of £2.50 p.w. and an upper limit of £4 p.w.**

1977   Pay increases restricted to 10 per cent.

1978   Attempt to restrict increases to 5 per cent.

Source: Clegg, 1979.

policies at the end of the 1970s, a small proportion of the population (about 10 per cent) were covered by minimum pay rates set by Wages Councils, until their eventual abolition in 1993. These Wages Councils covered a range of different industries such as catering and retail, although their power to enforce the minima that they set effectively diminished over the 1980s. This gradual erosion of the power of the councils is thought to have contributed to the increase in wage dispersion over the 1980s.[11]

4. *Trade unions.* The coverage of the work-force by trade unions has shrunk dramatically over the 1980s. Between 1980 and 1990, the proportion of the work-force belonging to a trade union fell from 58 per cent to 42 per cent. There is no a priori reason that unions should have an equalising rather than a disequalising effect on the distribution of wages. The presence of unions might result in higher wage differences between those who are covered by a union and those who are not, or between people covered by different union agreements. Empirically, however, higher union coverage has been associated with lower wage dispersion,[12] suggesting that the decline in union presence has contributed to the rising wage inequality over the last fifteen years.

SUMMARY AND CONCLUSIONS

This chapter has examined some of the major trends in the different sources of income. The main findings have been as follows:

1. Earnings currently make up the largest single source of household income; many of the trends in the income distribution that we described in Chapter 3 are therefore driven to a large extent by changes in the distribution of earnings; in particular, male wage inequality declined over the 1970s, when income inequality was falling, and rose over the 1980s, when the income distribution was becoming increasingly more unequal.

2. Changing patterns of participation in the work-force have also contributed to the trends in income inequality. Increased participation in the labour force over the 1970s came largely from married

---

[11] For example, see Machin and Manning (1994).
[12] See Gosling and Machin (1995) for the UK and Freeman (1993) and Card (1991) for the US.

women whose husbands were also in work; unemployment and falling male participation over the 1980s were mainly amongst households in which there were no other workers.

3. There has been considerable research in the UK and the US which finds that many of the changes in participation and the wage structure are related to a decline in the demand for unskilled workers and a rise in demand for skilled workers. Technological advances that have favoured skilled workers at the expense of the unskilled are one reason for this shift in demand. Institutional factors, such as government incomes policies and changes in the importance of trade unions, have also had an important role to play in explaining the changes in the wage structure that have taken place.

4. Earnings are by no means the whole explanation for the changes in the income distribution. The share of earnings in total income has fallen back considerably since the 1960s, and other sources of income have become correspondingly more important.

5. Social security is a major source of income for those at the bottom of the income distribution, and its share in total income has risen sharply over the 1970s and 1980s, as the number of people entitled to benefits has increased.

6. There have been a growing number of people reliant on incomes from self-employment, from private pensions, and from investments, particularly over the 1980s. These income sources have also played an increasingly important part in driving the changes in income inequality that we have seen.

6 Accounting for the Trends II: Demographic and Labour Market Changes

INTRODUCTION

This is the second chapter of three that explain the major changes in the income distribution over the last three decades. Which sorts of families have seen their average living standards move furthest ahead of the rest? Why have the different members of various population groups experienced such diverse outcomes over the various periods? These are the sorts of questions that this chapter is designed to answer.

The particular focus of the present chapter is the role of factors such as the changing demographic structure of the population and new patterns of family formation and dissolution, and also how the major changes in the labour market have affected different sorts of people. Chapter 7 goes on to focus on the role of the personal tax and benefit system in shaping trends in the income distribution.

For the purposes of this chapter, we use as our measure of income inequality the mean log deviation (MLD). As with the other measures of inequality that have been used so far, the MLD shows that inequality fell between the early and late 1960s, rose slightly between the late 1960s and early 1970s, fell back until the late 1970s, and then rose, at first steadily and then rapidly in the late 1980s.

The attraction of using the MLD is that, just as with half the squared coefficient of variation which we used in the last chapter, it is 'additively decomposable' in a relatively straightforward and intuitive way. What this means is that it is possible to split up total inequality on this measure into a range of component parts, all of which sum to the total inequality figure. In Chapter 5, we used one decomposable measure to divide overall inequality into contributions from different sources of income. The measure that we use in this chapter allows us to do decompositions of a slightly different sort. To give an example, it is possible to split the population by family type and to make statements of the form '*x* per cent of total inequality is attributable to inequality among single pensioners'.[1]

Table 6.1 applies this approach to pooled Family Expenditure Survey data at five-yearly intervals over the period 1961–63 to 1991–93.[2]

The top line of the table shows the aggregate MLD figure. When considering the contribution to total inequality made by different family types, it is possible to split this total figure into two principal components, namely a *between-group* component and a *within-group* component.

The between-group figure is a measure of the extent to which the differences in *average* income between the six different family-type categories are contributing to overall inequality. If all six family types had the same mean income (regardless of how that income was distributed within each group), then this figure would be zero.

The within-group figure measures how far income inequality within the six family-type groups is contributing to total income inequality. If all married pensioners had identical incomes, all single pensioners had identical incomes, and so on, then this figure would be zero, regardless of any differences in the average incomes of the various groups.

The remainder of Table 6.1 shows to what extent each family type is contributing to the total within-group inequality. Thus in 1991–93, couples with children are the main source of within-group inequality, contributing 72 points out of the total figure of 168. The breakdown of this total figure for within-group inequality is also shown graphically in Fig. 6.1.

The purpose of this decomposition, and of the further decompositions that we will undertake in this chapter, is to try to isolate exactly

---

[1] This general approach to examining trends in income inequality was first applied to UK data by Jenkins (1995). The present chapter seeks to develop this approach and to apply it to a longer run of data.

[2] The data have been adjusted in one important respect compared with those presented in the earlier chapters. For each broad source of income (earnings, self-employment, investments, etc.), the extreme values at the top of the distribution have been 'capped' at the ninety-ninth percentile and then household income has been recalculated by summing the adjusted components of income. The reason for this procedure is that some of the decompositions used later in the chapter can be distorted by the presence of even a small number of extreme values. As Table 6.1 shows, however, capping the extreme values does not significantly alter the general pattern of inequality over the three decades.

Table 6.1. Decomposition of total inequality by family type

	1961–63	1966–68	1971–73	1976–78	1981–83	1986–88	1991–93
Total inequality (= MLD × 1000)	**116**	**94**	**109**	**98**	**115**	**145**	**187**
Of which:							
Between family types	12	11	16	14	12	15	19
Within family types	104	83	93	84	103	130	168
Of which:							
Married pensioner	9	6	7	5	6	10	11
Single pensioner	9	7	7	5	5	6	8
Couple with children	43	35	39	39	47	55	72
Couple, no children	22	20	23	18	20	27	39
Single with children	3	3	3	3	3	4	6
Single, no children	18	13	15	14	21	28	33

Note: All values in this table have been multiplied by 1000 for ease of reading.

where the growth in inequality is occurring. Is it among people of working age or among pensioners? Among people with dependent children or without? These are the sorts of questions that decomposition analysis is able to answer.

There are, of course, an almost infinite number of different ways of splitting the population for the purposes of decomposition analysis. In part to mirror the descriptive analysis of the earlier chapters, we have chosen to split it into six groups that capture whether the family is above or below pension age, whether married or single, and whether there are any dependent children. This produces, in some cases, relatively homogeneous groups (such as lone parents, a large majority of whom are, at the outset of the 1990s, unwaged and receiving Income Support), but in other cases, some quite large and diverse groups (notably couples with children). Particularly in the case of the larger groups, we also perform further decompositions

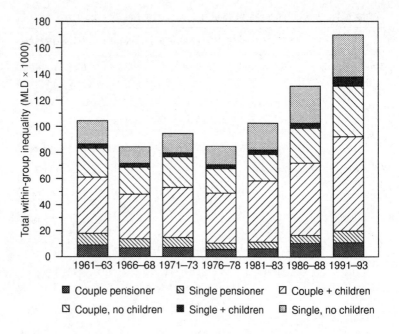

Fig. 6.1.   Total within-group inequality decomposed by family type

that take account, for example, of varying patterns of economic activity within the group (the number of workers in the family, etc.).

The remainder of this chapter, which is devoted to explaining the trends shown in Fig. 6.1, is structured as follows. We begin with a brief discussion of the results for inequality *between* family-type groups. The bulk of the chapter, however, is devoted to the results for within-group inequality, which, as Table 6.1 shows, accounts for the great majority of the overall picture. We consider each broad family type in turn and we undertake further decompositions to discover the factors that are producing the pattern of inequality shown in the table. A final section summarises the key conclusions from this analysis.

CHANGES IN INEQUALITY BETWEEN DIFFERENT
FAMILY-TYPE GROUPS

The values given in Table 6.1 for the between-group inequality show what the MLD would be if all members of each family-type group were given the mean income for that group. It therefore follows that changes in the between-group values directly reflect the divergence in the real mean incomes of the six family-type groups. These mean incomes are plotted in Fig. 6.2 for the period 1961–63 to 1991–93.

Clearly, all family types have, on average, enjoyed significant real income gains over the past three decades. However, it is also apparent that the differences between the mean incomes of the various groups have diverged, particularly during the 1980s. Thus in 1981–83, the mean income of the richest group (childless married non-pensioners who did particularly well during the 1980s) was around two-thirds higher than that of the poorest group (lone parents), whereas by 1991–

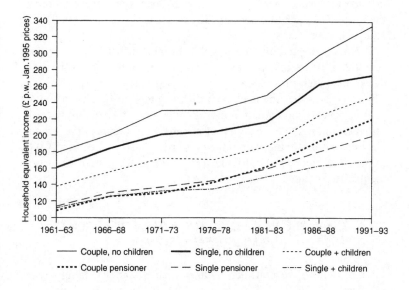

Fig. 6.2. Average real incomes by family type

93, the richest group was more than twice as well off as the poorest. This increase in dispersion explains the rise in between-group inequality over the 1980s shown in Table 6.1.

Figure 6.2 indicates that whilst the economy has grown significantly over the past three decades and especially over the 1980s, the fruits of that growth are not shared equally. Over the 1970s, those mainly dependent on social security benefits saw their living standards improve relative to the employed population (whose incomes were constrained by various forms of government pay policies). However, over the 1980s, the converse has been true. Whereas those in work have been able to command rapid real increases in their earnings, those dependent on benefits have seen much slower increases, especially since 1980 when the decision was taken to link the level of the Retirement Pension and related benefits only to the increase in retail prices.

A further notable feature of Fig. 6.2 is the dramatic improvement in the relative position of married pensioners. This group begins the period with a mean income similar to that of single pensioners and lone parents but ends it markedly better off than both. This is mainly because each successive cohort of married pensioners reaching retirement age has tended to have steadily larger entitlements to both state and private pensions.

It would, in principle, be possible to take each of the six family types identified here and to explain in great detail the factors that have determined their mean income over the past thirty years. However, as Table 6.1 indicates, the main contributor to total inequality in a given year is typically not variation in *mean* incomes between family types, but variation in the *distribution* of income within family-type groups. For this reason, we move on immediately to consider each family type in turn and to assess the driving forces behind the pattern of income inequality for each.

CHANGES IN INEQUALITY WITHIN DIFFERENT FAMILY-TYPE GROUPS

In this section, we consider in turn each of the six family types identified in Fig. 6.1. For both married and single pensioners, we have seen that overall inequality over the last three decades has displayed a U shape. For most non-pensioner groups, the pattern of inequality has been more akin to a 'squashed W', although with variations for particular sub-groups. The main reasons for these differences include the very different sources of income that the various groups receive and also the differential impact on particular groups of fluctuations over the past three decades in the level of economic activity.

Married Pensioners

Over the first twenty years under examination, the contribution of the married pensioner population to overall income inequality on this measure (the mean log deviation) declined slightly. It then rose quite sharply between 1981–83 and 1986–88, remaining at that level in 1991–93. The contribution of a particular group to overall income inequality is, however, derived from a combination of two factors—the extent of income inequality within the group and the size of the group. As Table 6.2 shows, in the earlier part of the period these two factors were working in opposite directions.

The first row of Table 6.2 shows the MLD figure just for married pensioners in each of the years. It shows that inequality *within* this section of society follows a U shape, falling markedly over the fifteen years from 1961–63 to 1976–78, but subsequently rising sharply, particularly in the second half of the 1980s. The contribution of these trends to overall population inequality is given by multiplying the figure for married pensioner inequality by the weight of this group in the population. In 1961–63, married pensioners made up 6.5 per cent of the population and so their contribution was nine points out of the UK total figure of 116 (see Table 6.1).

Table 6.2. Contribution of married pensioners to total inequality

	1961–63	1966–68	1971–73	1976–78	1981–83	1986–88	1991–93
(A) Unweighted sub-group MLD (× 1000)	136	85	88	66	72	108	112
(B) Population weight	6.5%	7.0%	7.8%	8.5%	8.7%	9.2%	9.3%
Contribution to total inequality (=A × B)	9	6	7	5	6	10	11

It is noticeable that on this measure, inequality within the married pensioner population in 1991–93 had not yet returned to its level of three decades ago. It follows that this group's (increased) contribution to overall inequality over the last three decades as a whole is explicable almost exclusively in terms of its increased share. It is interesting, however, to look at some of the sub-periods and in particular to examine the reasons for the U shape in married pensioner income inequality.

One way of examining the reasons for such trends is to look at the sources of income received by a given group of the population over the period. This enables us to assess whether the changes in total income inequality for the group in question reflect, for example, a change in earnings inequality rather than a change in inequality of pensions receipt.

To achieve this, we need to break down our total inequality measure further. In principle, our current measure (the mean log deviation) could be used, but the MLD does not in this case neatly decompose into readily interpretable components. Instead, we switch to using half the squared coefficient of variation as our measure of inequality. As in the last chapter, the contribution of each income source (earnings, pensions, investments, etc.) to total inequality among a given group can then be quantified and further explained in terms of the factor share of the income source, the correlation between that factor and total income, and the inequality with which that income source is distributed.

The first stage of this process is illustrated in Table 6.3, which shows total inequality for married pensioners on our new measure subdivided by income source. The decomposition undertaken here is equivalent to that shown in Fig. 5.5 in the last chapter, but is carried out just on the married pensioner population rather than the population as a whole.

A first interesting point to emerge from Table 6.3 is that although overall inequality for married pensioners on this measure still broadly exhibits a U shape, the trend between the late 1960s and early 1970s is actually slightly downwards on this measure rather than slightly upwards when using the MLD. This illustrates the important point that just because a given measure of inequality indicates a trend in one direction, it does not necessarily follow that all measures of inequality will move in the same direction, since each has its own properties and emphasises particular aspects of the distribution. In this case, the

Table 6.3. Decomposition of married pensioner income inequality by income source

	1961–63	1966–68	1971–73	1976–78	1981–83	1986–88	1991–93
Total inequality (= half the squared coefficient of variation × 1000)	**133**	**107**	**100**	**79**	**87**	**134**	**142**
Of which income from:							
Self-employment	25	15	10	10	11	18	12
Private pensions	14	13	21	20	28	44	51
Investment income	22	20	20	16	23	37	47
Earnings	88	73	63	43	34	39	31
Social security	−17	−15	−14	−10	−8	−5	−1
Other	1	1	0	0	0	0	1

MLD gives particular weight to households with low incomes, whereas the half the squared coefficient of variation measure of inequality is especially sensitive to very high incomes.

This does not, however, mean that it is never possible to make unambiguous statements about trends in inequality. Where, for example, the Lorenz curve for a given distribution in a particular year lies wholly within the curve for a previous year, then it is possible to say that on *all* standard inequality measures there has been a fall in inequality. It simply follows that in this case the Lorenz curves for married pensioners in the late 1960s and the early 1970s must cross.[3]

Returning to Table 6.3, it is apparent that the inequality of earnings plays a major part in total inequality at the start of the period, but that the importance of earnings inequality has declined quite rapidly by the mid-1970s. However, from the mid-1970s onwards, there is a noticeable rise in the contribution from private pensions and investments. Taken together, these two trends broadly account for the U shape in inequality among married pensioners over the period.

Table 6.3 also shows that social security benefits made a *negative* contribution to overall inequality among married pensioners, particularly in the earlier part of the period. This indicates that were the social security system to be abolished overnight, the distribution of married pensioner incomes would become more unequal. By 1991–93, however, this statement is only marginally true, since many better-off married pensioners are now receiving relatively significant state earnings-related pensions or are receiving two full basic pensions because both partners have a full record of lifetime National Insurance contributions.

As noted above, the reason for using this measure of inequality is that trends such as these can be more intuitively broken down and explained. Table 6.4 therefore focuses on just three of our sets of years (representing respectively the start, middle, and end of our period) and on four of the sources of income, to examine in more detail the reasons for the trends shown in Table 6.3.

Table 6.4 shows that each of the numbers given in Table 6.3 for the contribution of each factor to total inequality was in fact the product of four values. These were the value for the inequality with which that income source is distributed (square rooted) (A), the share of that income source in total household income (B), the correlation between

---

[3] For a more extensive discussion of this issue, see Jenkins (1991).

Table 6.4. Detailed decomposition of married pensioner income inequality

Contribution of:	1961–63	1976–78	1991–93
Private pensions	**14**	**20**	**51**
(= A × B × C × D)			
(A) Within-source inequality (square root × 1000)	1193	1049	889
(B) Factor share	0.12	0.14	0.24
(C) Correlation with total income	0.26	0.48	0.64
Investment income	**22**	**16**	**47**
(= A × B × C × D)			
(A) Within-source inequality (square root × 1000)	1860	1456	1211
(B) Factor share	0.08	0.08	0.15
(C) Correlation with total income	0.40	0.50	0.70
Earnings	**88**	**43**	**31**
(= A × B × C × D)			
(A) Within-source inequality (square root × 1000)	1112	1499	1982
(B) Factor share	0.32	0.17	0.09
(C) Correlation with total income	0.67	0.60	0.46
Social security	**−17**	**−10**	**−1**
(= A × B × C × D)			
(A) Within-source inequality (square root × 1000)	331	187	193
(B) Factor share	0.42	0.59	0.50
(C) Correlation with total income	−0.34	−0.32	−0.02
(D) Total inequality (all sources, square root)	0.36	0.28	0.38

a given income source and total household income (C), and the square root of the figure for inequality of total income (D). (For a discussion of this breakdown, see Chapter 5.) This split enables us to provide an interpretation of the trends shown in Table 6.3.

Consider first *private pension* income. The trend in Value A for this income source shows that the distribution of receipt of private pension income, whilst still being very unequal, became less so between 1961–63 and 1991–93. This partly reflects the fact that more people now receive income from private pensions than was the case over the 1960s; when more people are in receipt of a particular income source, there are fewer 'zero values' recorded amongst the population, and so the measured inequality in the distribution of that source falls. In part, this fall in the value for A may also reflect the effects of high inflation in eroding the real value of inadequately-indexed pensions in pay-ment. Value B shows that the share of private pension income in total married pensioner income has risen from 12 per cent in the early 1960s to 24 per cent in the early 1990s. Value C indicates that private pensions have steadily become more positively correlated with total income over the period. This reflects the facts that better-off married pensioners are increasingly likely to have private pension income and that average amounts of private pension income among recipients in better-off households have risen. Finally, as we have seen, overall income inequality for married pensioners (Value D, at the bottom of Table 6.4) fell sharply between the early 1960s and mid-1970s and then rose sharply between the mid-1970s and the early 1990s.

As regards private pension income, therefore, we would have expected some increased contribution to total inequality simply because the factor share has risen—in other words, because it has become a more important income source for this group. However, what our results also show is that higher private pensions are increas-ingly likely to be associated with higher total incomes. Taken together, these factors account for the rising contribution to inequality of this income source over the period, and in particular since the mid-1970s.

Considering next the contribution of *investment income*, a first interesting trend is that the distribution of this income source has actually become more equal (with inequality falling from 1860 to 1211 over the three decades), reflecting again the fact that there are now more married pensioners who receive investment incomes. How-ever, it remains very unequally distributed and so its contribution to total inequality is heavily dependent on how much investment income is being received in aggregate. The factor share row (B) indicates a steady share between the early 1960s and mid-1970s, but a sharp rise thereafter, probably reflecting the greater affluence of this group. It is also the case that investment income is increasingly positively correlated with total household income for this group. As overall incomes became more compressed in the first part of the period, so the contribution of investment income fell, whilst as inequality subsequently rose, so too did the investment incomes of the better off.

The contribution of *earnings* and also of self-employment income (not shown in Table 6.4) declined considerably between the 1960s and 1970s, before stabilising during the 1980s. The strongest effect may be seen in the declining factor share of this income source. In 1961–63, earnings formed almost a third of total income for married pensioners. By 1976–78, the proportion was down to 17 per cent, and by the early 1990s down further to 9 per cent. This change alone accounts for the fall in contribution of this source in the first part of our period. The contribution of earnings would have fallen further between the mid-1970s and early 1990s but for the fact that those earnings that are received are now much more unequally distributed (row A). This partly reflects the fact that fewer pensioners receive any income from earnings at all; it is also part of the more general economy-wide phenomenon of growing wage inequality amongst those in work.

Looking finally at *social security*, some rather interesting trends are visible. First, the negative sign on the correlation coefficient (C) indicates that, in general, those married pensioners with higher household incomes tend to receive lower social security benefits. This is to be expected since certain social security benefits such as Supplementary Benefit (the precursor to Income Support) and Rent Rebates were explicitly restricted to those on low incomes. What is perhaps more surprising is that the link between higher social security benefits and lower total incomes actually became much weaker over the 1980s for this group. The main reason for this is likely to be the growth in receipt of basic state pensions by married women. A rise in female labour market participation in earlier decades has now begun to feed through into higher state pension entitlements for women, and these will in many cases be women whose husbands already have significant pension income.

The factor share of social security rose over the 1970s, partly because of the introduction of a national system of Rent Rebates, but fell back slightly over the 1980s despite the growth of the State Earnings-Related Pension Scheme (SERPS). A major reason for this is that from 1980 onwards, the basic state pension was uprated each year only in line with increases in retail prices. Since pensioner incomes, and in particular private pensions and investments, rose at a much faster rate, this contributed to a fall in the factor share of social security.

Social security benefits also became less unequally distributed amongst the group (row A), again probably reflecting the growth in housing benefits over the 1970s. This reason, coupled with the falling negative correlation between benefit receipt and total income, explains the declining equalising effect of social security benefits for this group.

Summary

1. The contribution of married pensioners to UK income inequality declined somewhat from the early 1960s to the early 1980s, but rose markedly in the late 1980s and has remained at this higher level in the early 1990s.
2. Inequality amongst married pensioners displays a U shape over the last three decades, but the effect of this on overall inequality is slightly masked by a gradual growth in the size of the group.
3. The fall in inequality amongst this group between the late 1960s and late 1970s is attributable mainly to a fall in the contribution to inequality from earned income. This arose mainly from a fall in employment levels among this group.
4. The rise in inequality between the mid-1970s and early 1990s was principally attributable to an increase in the contribution from private pensions and investment income. Both sources formed an increasing share of total married pensioner income and are increasingly likely to be accompanied by significant incomes from other sources.

Single Pensioners

Single men aged 65 or over and single women aged 60 or over currently form around 8 per cent of the UK household population. Table 6.5 shows how the contribution of this group to total UK income inequality (as summarised in Table 6.1) is constructed.

It is immediately apparent from Table 6.5 that the U shape in income inequality that we saw earlier for married pensioners is also a marked feature of the trend in income inequality among single

Table 6.5. Contribution of single pensioners to total inequality

	1961–63	1966–68	1971–73	1976–78	1981–83	1986–88	1991–93
(A) Unweighted sub-group MLD (× 1000)	133	95	89	64	59	76	100
(B) Population weight[a]	7.0%	7.1%	7.4%	7.6%	8.1%	7.9%	7.8%
Contribution to total inequality (= A × B)	9	7	7	5	5	6	8

[a] The population in question here is the *household* population, and it thus excludes those living in institutions such as residential care homes. The rise in the numbers of single elderly people living in such homes largely explains the slight fall in the numbers of single pensioners shown in this table.

pensioners. The next step in our analysis is to decompose this result to see how far the reasons for this overall pattern are the same as for married pensioners. Summary results are presented in Table 6.6. This decomposition by income source again switches to using half the squared coefficient of variation as the measure of inequality.

In many ways, these results resemble those for married pensioners presented in Table 6.3. The fall in the contribution of earnings inequality is a major factor in the general fall in single pensioner inequality over the 1960s and 1970s, whilst from the mid-1970s onwards, both private pension incomes and investment incomes contribute increasing amounts to overall inequality.[4]

There are, however, some differences between the results for single pensioners and those for married pensioners. In particular, whilst the two groups start the period with very similar levels of within-group inequality, by the end it is amongst married pensioners that there is clearly most inequality.

It is also noticeable that social security has more of an equalising effect among the single pensioner population than amongst married pensioners, although in both cases this role is much diminished compared with earlier decades. In part, this reflects the fact that older

[4] For a more detailed discussion of trends in the pensioner income distribution, see Johnson and Stears (1995) and Hancock and Weir (1994).

Table 6.6. Decomposition of single pensioner income inequality by income source

	1961–63	1966–68	1971–73	1976–78	1981–83	1986–88	1991–93
Total inequality (= half the squared coefficient of variation × 1000)	**136**	**116**	**113**	**71**	**72**	**106**	**123**
Of which income from:							
Self-employment	11	13	13	8	6	11	6
Private pensions	8	9	14	13	19	34	39
Investment income	26	17	18	12	16	28	40
Earnings	113	97	88	51	42	42	41
Social security	–23	–20	–19	–13	–11	–10	–5
Other	1	0	0	0	0	0	0

pensioners (who are mainly single) have begun to receive relatively generous treatment through the means-tested benefit system.

Summary

1. Inequality among single pensioners exhibits a similar U shape to that for married pensioners. A downward trend in earnings inequality is followed by an upward trend in the contribution from private pensions and investments.
2. Although inequality amongst single pensioners has risen since the early 1980s, this group is still less unequal than married pensioners.

Couples with Dependent Children

Married or cohabiting couples with dependent children form the largest of our six family-type categories, comprising around 37 per cent of the UK population in 1991–93. The contribution of this group to total inequality is shown in Table 6.7.

Table 6.7 shows that whilst total inequality amongst couples with children fell somewhat over the 1960s, it has since risen steeply and steadily. This trend would have had an even more pronounced effect on total UK inequality were it not for the fact that this group has diminished sharply in size. This fall is in part attributable to a growth in lone parenthood, which in turn reflects both rising divorce rates and a higher incidence of childbirth among younger never-married mothers.

Table 6.7. Contribution of couples with children to total inequality

	1961–63	1966–68	1971–73	1976–78	1981–83	1986–88	1991–93
(A) Unweighted sub-group MLD (× 1000)	95	77	86	87	111	142	192
(B) Population weight	45.2%	45.5%	45.6%	44.9%	42.7%	38.9%	37.3%
Contribution to total inequality (= A × B)	43	35	39	39	47	55	72

A comparison of Table 6.7 with the corresponding tables for married and single pensioners also indicates that whilst, in the early 1960s, pensioner incomes were much more unevenly distributed than those of working-age couples with children, by the early 1990s, the converse is very much the case. Thus in 1961–63, the MLD of 95 for working-age couples with children compared with a figure of 136 for married pensioners and 133 for single pensioners. The corresponding figures for 1991–93 were 192, 112, and 100 respectively.

What are the reasons for the sharp rise in inequality amongst working-age couples with children? In order to answer this question, we begin by further subdividing this rather large group according to whether the household of which the couple is a member contains zero, one, or two workers.[5] In this case, a worker is defined as an employee (either full-time or part-time) or someone who is self-employed. Figure 6.3 shows how the relative size of these groups has changed over the last three decades, whilst Table 6.8 shows how the total inequality of couples with children is divided between these three categories.

Figure 6.3 indicates that the decline in size of this group has occurred principally amongst the 'traditional' one-earner family. The proportion of no-earner couples with children has risen markedly, mainly because of a rise in unemployment, whilst the proportion of two-earner families also rose particularly during the 1970s mainly due to a rise in the rate of female labour market participation, principally in part-time jobs. This combination of higher male unemployment and increased female employment, coupled with the general decline in the number of families with dependent children, has meant that members of one-earner families with children now make up less than one-eighth of the total UK population.

The first row of Table 6.8 repeats the results for the contribution to total inequality of couples with children, whilst the following rows show how each total is broken down according to within- and between-group inequality when this group is further divided according to the number of workers in the household. It should be noted that these values are all directly comparable with those in Table 6.1. In other words, of the total UK inequality (based on the

---

[5] Strictly speaking, the final category should be 'two or more', since couples with children may be part of a larger household where there are more than two workers.

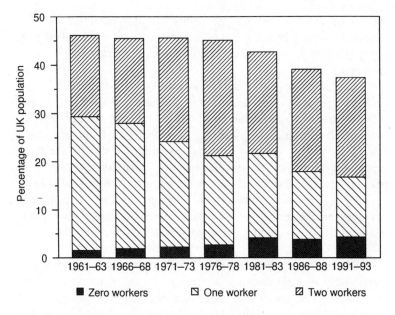

Fig. 6.3. Couples with children as a percentage of UK population, by number of workers in household

Table 6.8. Couples with children: decomposition by number of workers in household of contribution to inequality

	1961–63	1966–68	1971–73	1976–78	1981–83	1986–88	1991–93
Contribution to total inequality (= MLD × 1000)	43	35	39	39	47	55	72
Of which:							
Between economic types	6	5	6	5	8	10	13
Within economic types	37	30	33	34	39	45	59
Of which:							
No workers	5	5	5	4	4	3	5
One worker	23	17	16	16	18	19	26
Two workers	8	9	11	14	17	23	28

MLD) of 116 in 1961–63, 23 points are accounted for by one-earner couples with children, whilst of the total MLD of 187 in 1991–93, just 5 points are accounted for by no-earner couples with children.

From Table 6.8, it is clear that amongst couples with children, the relative contribution of the different economic types has changed considerably. At the beginning of the period, one-worker families were by far the most important group, whereas by the late 1980s and early 1990s, two-earner couples had become slightly the most important contributors to overall inequality. In part, this reflects group size effects, as shown in Fig. 6.3. However, more detailed analysis by income source (using the methods already applied to pensioner incomes) indicates that for both these groups there has been a dramatic increase in earnings inequality which is contributing to the overall trend.

What factors can account for this growth in earnings inequality, and particularly in male earnings inequality? A first important point to remember is that for the purpose of this study, earnings are measured after the deduction of direct taxes. Given that income tax was cut quite significantly over the 1980s, particularly for high earners, this would of itself tend to lead to a rise in earnings inequality even with an unchanged pre-tax distribution. Johnson and Webb (1993) argue that taking just the period 1979–88, had the 1979 direct tax system been retained, the rise in total UK income inequality would have been halved. The role of the tax and benefit system is considered more fully in Chapter 7.

There is, however, considerable evidence that the distribution of pre-tax earnings has also widened substantially since the late 1970s. The reasons for this very important change have been discussed more fully in Chapter 5.

The rise in inequality in self-employment incomes that has occurred among working couples with children is also worthy of note. In part, this reflects the general rise in self-employment, which formed 9 per cent of the household income of two-earner couples in 1976–78 and 15 per cent in 1991–93. Those with income from self-employment are also increasingly likely to have income from other sources (such as earnings), and this combination tends to exacerbate inequalities. The incomes and circumstances of the self-employed are, however, an area where comparatively little is known and where existing household surveys contain insufficient detail to establish a clear picture.[6]

Summary

1. Inequality amongst couples with children has risen sharply since the late 1970s. Within-group inequality amongst this group is now markedly greater than for single or married pensioners, whereas at the start of the 1960s the converse was the case.
2. The size of the group has shrunk and its composition has changed. Higher unemployment has produced more no-earner couples, whilst a rise in female part-time employment has led to a rise in the number of two-earner couples. Members of one-earner couples with children have fallen from 28 per cent of the population in the early 1960s to 12 per cent in the early 1990s.
3. The rise in inequality amongst working couples with children is largely attributable to an increase in the inequality of earnings and self-employment incomes within the group.

Couples with No Dependent Children

Married or cohabiting couples with no dependent children may be classified into two broad groups—older couples who have never had children or whose children have now grown up, and younger couples who currently do not have children though who may have later in life. For brevity, we refer to both groups as 'childless couples'. The fact that this category of families contains two distinct groups may be seen in Fig. 6.4, which provides an age distribution of the husbands (or male partners) in non-pensioner couples with and without dependent children, based on the 1991–93 data.

In the analysis that follows, we split this group between the under- and over-40s in order to examine their different circumstances in more detail. We begin, however, in the usual way by showing in Table 6.9 how the contribution of all childless couples to aggregate inequality is derived.

The unweighted figures for the MLD of childless couples indicates that for this group, inequality was relatively stable over the late 1960s

---

[6] See Meager, Court, and Moralee (1996) for a recent survey of evidence on the living standards of the self-employed.

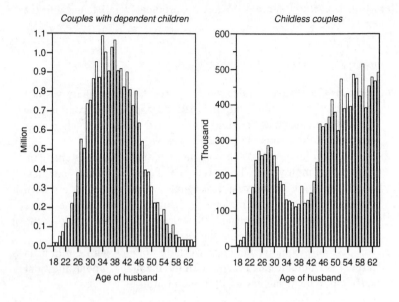

Fig. 6.4. Age distribution of men in couples with dependent children and in childless couples, 1991–93

Table 6.9. Contribution of childless couples to total inequality

	1961–63	1966–68	1971–73	1976–78	1981–83	1986–88	1991–93
(A) Unweighted sub-group MLD (× 1000)	99	89	107	91	110	145	171
(B) Population weight[a]	22.4%	22.2%	21.8%	20.2%	18.6%	18.7%	22.7%
Contribution to total inequality (= A × B)	22	20	23	18	20	27	39

[a] To be consistent with our sample definition of a childless couple (which includes cohabiting couples), our census-based estimates of the UK population (which identify legal marriage only) should have been adjusted. This has only been done for the 1991–93 data. This suggests that the true numbers in this group have probably remained at about the 22 per cent mark, but that the proportion who are legally married has been falling. Regrettably, reliable estimates of the extent of cohabitation are not available for the whole period.

and 1970s, but then takes off in dramatic fashion over the 1980s. In order to investigate these trends in more detail, we split the group into the under-40s and the over-40s (or strictly the '40s and over') and examine in Table 6.10 how far each has contributed to the overall trend.

Considering first the under-40s, it is noticeable that their incomes are relatively evenly distributed as compared with other working-age groups. Thus their within-group MLD of 115 in 1991–93 compares with the figure of 186 for childless over-40s and 192 for couples with children as a whole. The population weights, although distorted by the discontinuity noted in Table 6.9, probably indicate a gradual increase in size of this group, certainly over the period as a whole. One reason for this is a trend for those women who eventually have children to have their first child at a later age, often giving the woman more opportunity to become more established in a career.

The group of childless under-40s is also one of the richest that we identify, with a mean income in 1991–93 of around £410 per week

Table 6.10. Contribution of childless couples to total inequality, split by under- / over-40

Contribution to total inequality of:	1961–63	1966–68	1971–73	1976–78	1981–83	1986–88	1991–93
All childless couples	22	20	23	18	20	27	39
Of which:							
Under-40s							
(A) Unweighted sub-group MLD (× 1000)	64	48	90	81	85	111	115
(B) Population weight	4.5%	4.4%	5.0%	5.4%	4.8%	4.8%	6.4%
Total (= A × B)	3	2	4	4	4	5	7
Over-40s							
(A) Unweighted sub-group MLD (× 1000)	105	96	108	89	112	149	186
(B) Population weight	17.9%	17.9%	16.8%	14.8%	13.8%	13.9%	16.3%
Total (= A × B)	19	17	18	13	16	21	30

Note: The between-group inequalities for the under- / over-40 split are negligible and are excluded from the table.

compared with a population mean of roughly £275. This is partly because they do not have to meet the costs of children (which would be reflected in a lower 'equivalent' income)[7] but mainly because the absence of children makes it possible in most cases for both partners to work. Thus in 1991–93, 82 per cent of childless couples under 40 had both partners working and only 4 per cent had neither partner in work.

Amongst the over-40s, inequality is markedly higher, reflecting the greater diversity of this group. Whilst around three in five childless over-40s were in two-earner households in 1991–93, around 15 per cent had no earner. This latter result often reflects early retirement, whether involuntary (through sickness or redundancy) or voluntary, and perhaps accompanied by a sizeable private pension.

The fluctuations in inequality within this group have been most marked since the early 1970s. Table 6.11 shows how these fluctuations are broken down by income source, again switching to using half the squared coefficient of variation as our measure of inequality. In this case, the fluctuations in that measure mirror quite closely the MLD-based results shown in Table 6.10.

Considering first the earlier part of the period, it is clear that the major factor in the fall in inequality between the early and late 1970s was a decline in earnings inequality. One of the main reasons for this is likely to have been the effect of the various government pay policies implemented during the mid-1970s. In a number of cases, the policy required that pay increases should contain a flat-rate element (worth proportionately more to lower earners) or some cap on the absolute increases of higher earners, both of which would tend to reduce earnings inequality.

Looking at changes during the 1980s, as well as the by now familiar increase in earnings inequality, we see that investment and self-employment incomes are also contributing significantly to the increased inequality. It is worth remembering, however, that the incomes of the self-employed are particularly difficult to measure using the information gathered in all-purpose household surveys such as the Family Expenditure Survey. It may well be the case that some of the very low (or negative) self-employment income figures that have driven up measured inequality amongst older

[7] See Banks, Blundell, and Lewbel (1992) on life-cycle equivalence scales.

Table 6.11. Decomposition of income inequality by income source: childless couples over 40

	1971–73	1976–78	1981–83	1986–88	1991–93
Total inequality (= half the squared coefficient of variation × 1000)	**88**	**66**	**86**	**119**	**137**
Of which income from:					
Self-employment	6	4	6	16	14
Private pensions	0	0	1	3	4
Investment income	5	3	5	10	14
Earnings	82	66	85	102	117
Social security	−5	−7	−11	−12	−12
Other	0	0	0	0	0

childless couples are not necessarily an accurate reflection of their true living standards.

Summary

1. Members of childless couples of working age comprise slightly more than one-fifth of the UK population and contribute a similar proportion to overall income inequality.
2. Within the category of childless couples, it is possible to identify two distinct groups—younger couples (married or cohabiting) who may be yet to start a family, and older couples whose children may have grown up or who may never have had children.
3. The younger childless couples are, on average, a very prosperous group with an inequality level that is relatively low compared with that of other groups of working age. The vast majority are two-earner couples.
4. Older childless couples are a more diffuse group with greater within-group income inequality; their experiences of inequality broadly mirror those of the whole population, with falls in earnings inequality in the 1970s, followed by rises in the inequality of earnings, investment income, and self-employment income over the 1980s.

Single People with Dependent Children (Lone Parents)

Single non-pensioners with dependent children, or lone parents, comprise the smallest of our six broad family groups, but also the fastest growing. In 1991–93, around 3.8 million adults and children, or just under 7 per cent of the population, were members of lone-parent families. This compares with around 2.4 per cent in 1961–63. Table 6.12 summarises the trends in the size of this group together with its contribution to overall income inequality.

The contribution of this group to overall UK income inequality in any given year is clearly very small. In part, this reflects the small (though rapidly-rising) population share of the group, but also the fact that, amongst lone parents, incomes are relatively evenly distributed. In particular, around 70 per cent of lone parents are currently in receipt of means-tested Income Support which means that the majority of lone parents have a broadly similar living standard.

It must be stressed that this does not necessarily imply that the growth in lone parenthood has not had an impact on UK income inequality. For example, if the number of young single women becoming mothers had risen less quickly, there would have been more single childless women who might have been in employment rather than in receipt of social security benefits as lone parents. It is, however, difficult to say a priori what effect this alternative pattern would have had on overall levels of inequality. To make a judgement on this issue requires a view on what society and incomes would have looked like had the growth in lone parenthood not taken place, and this would inevitably be highly speculative.

Table 6.12. Contribution of lone parents to total inequality

	1961–63	1966–68	1971–73	1976–78	1981–83	1986–88	1991–93
(A) Unweighted sub-group MLD (× 1000)	106	92	77	65	66	86	94
(B) Population weight	2.4%	2.7%	3.3%	4.1%	4.5%	5.0%	6.8%
Contribution to total inequality (= A × B)	3	3	3	3	3	4	6

There have, however, been some quite dramatic changes both in the composition of the lone parent population and in the economic position of households containing lone parents. To illustrate this point, Fig. 6.5 shows the proportion of lone parents in each of the sets of years under examination, who were members of households with zero, one, or two workers.

The most striking feature of Fig. 6.5 is the remorseless rise in the proportion of lone parents living in households where there is no one in employment. By 1991–93, around 58 per cent of lone parents were in this position, whilst a further third of the lone parents in one-worker households were only in part-time employment and needed Income Support to top up their income.

Although divorcees (together with separated mothers) remained the most numerous group of lone parents, single (that is, never-married) lone parents were the fastest-rising group. These individuals, almost exclusively women, tended on average to be much younger than their divorced counterparts and to have younger children. Partly as a consequence of this, relatively few of them are in any kind of paid employment.

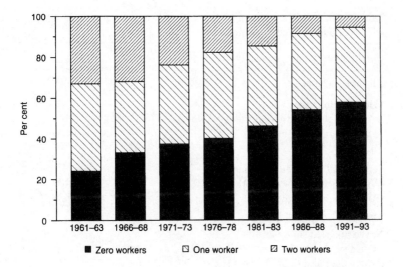

Fig. 6.5. Number of workers in households containing a lone parent

Also very striking from Fig. 6.5 is the fact that three decades ago, around a third of lone parents were living in households with two (or more) workers. This fact illustrates the major change that has taken place within the lone-parent population. In the 1960s and particularly the 1970s, the dominant and fastest-rising group of lone parents was divorcees. Many of these women would have had older children and so found it relatively easy to maintain some kind of paid employment. It appears that many of them also shared a household with another wage-earner, typically a grown-up child or other family member.

Unfortunately, our survey data do not allow us to identify the detailed marital status of lone parents until 1991–93. However, even the results for that set of years alone, presented in Table 6.13, illustrate some of the general points made above.

Two aspects of the foregoing discussion are confirmed in Table 6.13. The first is the relative size of the different groups. Widowed lone parents and their dependent children together form less than half of one per cent of the population compared with around 4 per cent for separated or divorced parents with dependent children (split roughly equally between those two categories) and around 2 per cent for never-married lone parents. The second is the extent of inequality

Table 6.13. Contribution of lone parents to total inequality, split by widowed/divorced/never-married

Contribution to total inequality of:	1991–93
All lone parents	**6**
Widowed lone parents	
(A) Unweighted sub-group MLD ($\times$ 1000)	141
(B) Population weight	0.3%
Total (= A $\times$ B)	**0**
Separated / divorced lone parents	
(A) Unweighted sub-group MLD ($\times$ 1000)	100
(B) Population weight	4.2%
Total (= A $\times$ B)	**4**
Never-married lone parents	
(A) Unweighted sub-group MLD ($\times$ 1000)	68
(B) Population weight	2.1%
Total (= A $\times$ B)	**1**

within each of these groups. The within-group MLD for never-married lone parents is just 68 points, indicating a high concentration of incomes. The corresponding figure for divorced and separated parents is much higher at 100 but this is still relatively modest compared with other family types.

Summary

1. In the early 1990s, adults and children in lone-parent families formed just under 7 per cent of the UK population, a rise of almost threefold since the early 1960s.
2. The contribution of this group to overall UK income inequality in any given year is very small, partly because of the small group size but also because of the relative homogeneity of lone parents' incomes.
3. In the 1960s, the largest single group of lone parents was divorcees who typically had older children and were in some form of paid employment. They also often shared a household with another wage-earner.
4. By the 1990s, the fastest-growing group of lone parents was never-married mothers, of whom a high proportion had no paid employment. The incomes of this group are highly concentrated around basic Income Support levels.

Single People with No Dependent Children

The final broad population group is single non-pensioners without dependent children. This predominantly younger group includes young adults still living in their parents' home and students, as well as single people of all ages who are living independently. Table 6.14 shows the contribution of single childless non-pensioners to total UK income inequality in our standard format.

The broad trend in within-group inequality revealed by Table 6.14 for the single childless group is similar to that for the population as a whole. It is, however, more illuminating to break down this rather large group according to the number of workers in the household. A summary of the numbers involved is shown in Table 6.15.

The first clear trend apparent from Table 6.15 is that growing numbers of single people are members of households where there is no earner. The figure of 3.3 per cent for 1991–93 represents just under

Table 6.14. Contribution of single childless non-pensioners to total inequality

	1961–63	1966–68	1971–73	1976–78	1981–83	1986–88	1991–93
(A) Unweighted sub-group MLD (× 1000)	108	86	104	94	122	137	206
(B) Population weight[a]	16.5%	15.6%	14.3%	15.1%	17.3%	20.4%	16.1%
Contribution to total inequality (= A × B)	18	13	15	14	21	28	33

[a] This series suffers from the same discontinuity as that for childless couples. The estimates for the 1980s in particular are likely to include some members of childless cohabiting couples who were coded as single in the census data on which the population estimates are based.

Table 6.15. Single childless people, by number of workers in household

(Percentage of UK population)	1961–63	1971–73	1981–83	1991–93
No workers	0.9%	1.0%	2.5%	3.3%
One worker	4.3%	4.3%	5.6%	6.1%
Two workers	11.4%	9.0%	9.2%	6.6%

2 million individuals, a more than threefold increase since 1961–63. Also notable is the decline in the proportions living in households with two (or more) workers. By definition, the single childless adults in this situation must be sharing a household, typically either with their parents or with contemporaries. A general trend for young adults to leave home at an earlier stage is likely to have been the principal reason for this fall.

Of these groups, it is the single people in *no-worker households* whose situation is the most diverse. As Fig. 6.6 indicates, this group incorporates young and old, and its composition has changed markedly over the last three decades.

Three broad trends are apparent from Fig. 6.6.

1. The number of economically inactive single people has risen markedly over the last three decades. To a large extent, this reflects the growing numbers of people in full-time education, but the sub-group also includes those who have withdrawn from the active

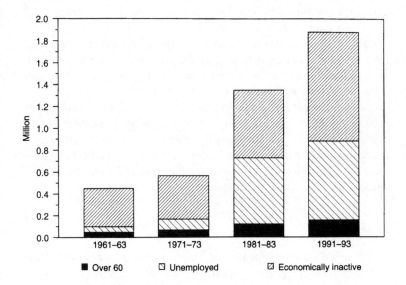

Fig. 6.6. Economic status of single people in no-worker households

labour market, either due to sickness or because they have given up looking for work.

2. The number of unemployed single people (defined as those who said they were 'seeking work') stood at around 725 000 in 1991–93 compared with less than 100 000 in 1961–63. Although this number varies considerably over the economic cycle, the numbers unemployed at the economic high point of successive cycles is now considerably higher than they were over the 1960s.

3. The number of single men aged 60 to 64 no longer in employment has risen significantly to stand at around 150 000 in 1991–93.

The diversity of this group is reflected in a relatively high absolute figure for within-group income inequality (a MLD of 219 in 1991–93 compared with the national figure of 187). However, because single people in no-earner households still represent a fairly small proportion of the total population, the contribution of this group to overall inequality is still relatively modest.

As regards the larger group of single people in *one-worker households*, an interesting insight into its contribution to total inequality may be gained by examining the sources of income of this group. These results (based on half the squared coefficient of variation, as previously) are presented in Table 6.16.

Table 6.16 indicates that the overall trend in inequality for the one-earner single childless is largely driven by the contribution of earnings inequality, and the reasons for the trends in earnings inequality have already been discussed.

The final sub-group, namely the single childless in *two-worker households*, has the lowest within-group inequality of the three groups of single childless. This indicates that those in two-worker households are a relatively homogeneous group, although, as Table 6.15 indicated, this combination of domestic and economic circumstances is becoming increasingly uncommon.

Summary

1. Single childless non-pensioners make up around one-sixth of the UK population. Their aggregate experience of trends in inequality over the past three decades is similar to that of the UK as a whole.
2. Single people in no-worker households are a rapidly-growing group, principally because of rising unemployment and increased participation in full-time education. The financial circumstances of this group, which also includes the early-retired, remain relatively diverse, although less so than in earlier decades.

Table 6.16. Decomposition of income inequality by income source: single people in one-worker households

	1961–63	1971–73	1981–83	1991–93
Total inequality (= half the squared coefficient of variation × 1000)	**116**	**94**	**101**	**141**
Of which income from:				
Self-employment	4	6	4	6
Private pensions	4	2	1	2
Investment income	16	5	6	14
Earnings	99	90	104	133
Social security	−8	−9	−14	−12
Other	1	0	0	−1

3. One-worker households containing single people have experienced a marked rise in income inequality in recent years, with a particular rise in the inequality of earnings.
4. Single people living in two-worker households form a diminishing group. This group has relatively uniform living standards.

CONCLUSIONS

The total level of income inequality in a given country arises from a combination of two factors—differences in the average incomes of different population groups and differences in the distribution of income within population groups. In the UK over the last three decades, and treating the population as comprising six broad demographic groups, the main determinant of the level of inequality in any year has been inequalities in the distribution of income *within* different family groups rather than differences in average incomes *between* different family groups.

The trend in UK inequality over the last three decades has been driven largely by the experiences of the population of working age. Whilst this is to some extent inevitable, given the relatively large proportion of the population accounted for by this group, fluctuations in earnings inequality have clearly played a central role in determining the overall trend in income inequality.

For the period 1961–63 to 1991–93 taken as a whole, the within-group inequality of working-age couples with children and of single childless people has mirrored the 'flattened-W' shape characteristic of the population as a whole. For pensioners by contrast, and to a lesser extent for lone parents, inequality has been 'U-shaped'.

As regards individual family groups, the main conclusions are:

1. *Married pensioners.* The initial drop in overall income inequality is driven by a fall in inequality from earnings, whilst the subsequent rise is attributable to a rise in the contribution of private pensions and investment income.
2. *Single pensioners.* Like married pensioners, this group has a U-shaped experience of inequality. The main difference is that the incomes of single pensioners end the period less unequally distributed than those of their married counterparts.
3. *Couples with children.* This group sees a large rise in aggregate inequality over the period, starting off less unequal than pensioners

but ending up markedly more so. The rise in inequality among the diminishing number of one-earner families mainly reflects the rise in male earnings inequality, whilst the rise among two-earner families also reflects increased inequality of self-employment income.

4. *Couples without children*. The younger childless couples are a prosperous group with relatively homogeneous living standards; the older couples are a more diverse group whose experience of trends in inequality mirrors that of the population as a whole.

5. *Single with children*. Lone parents make a small contribution to overall inequality in any given year partly because of small group size but also because of a relatively compact within-group distribution of income. A growing proportion of lone parents are never-married mothers dependent on Income Support and hence with very similar living standards. The incomes of separated and divorced lone parents are, however, more diverse and appear to have become increasingly so over the last fifteen years.

6. *Single with no children*. This predominantly young group contains a relatively wide diversity of individual circumstances, and the group's proportionate contribution to total inequality has risen in recent years. Single people in no-worker households have become increasingly numerous, but the main rise in inequality has been among one-worker households.

7   Accounting for the Trends III: The Tax and Benefit System

INTRODUCTION

Because we are interested in people's living standards, we are interested in their net income—that is, their income after they have received any social security benefits and paid any taxes on their income. On top of all the changes to original or market income that occur over time are the changes made by government to the tax and benefit system. Understanding the impact of taxes and benefits is a crucial part in understanding overall trends in the income distribution.

To analyse the impact of changes in the tax and benefit system, it is important to begin by considering how the current system operates and how it affects households of different types. It will then be possible to see more clearly why particular changes to the system have had the effects that they have had. The initial analyses, then, describe the impact of the tax and benefit system on household incomes in 1991–93, the latest three years for which micro-data are available, before moving on to look at changes in taxes and benefits since the early 1960s. But before going on to do this, it is worth considering the alternative strategies for examining this question.

ANALYSING EFFECTS OF TAXES AND BENEFITS

The most obvious way to measure changes in the impact of the tax system is to look at how much tax different types of household are actually paying in a given year and to examine how that pattern changes over time. This approach formed the basis for a study by Jenkins (1995), who concluded that changes in tax payments explained relatively little of the growth in post-tax income inequality over the 1980s. We refer to this approach as the 'actual payments' approach.

One problem with the 'actual payments' approach is that the amount of tax a household pays is a consequence of (at least) two factors—the amount of (pre-tax) income they have, and the rates of tax that are applied to that income. It follows that looking only at the amounts of tax paid makes it impossible to distinguish between changes in the tax structure and changes in the distribution of pre-tax incomes. An alternative approach (adopted by Johnson and Webb (1993)) is to hold the pre-tax income distribution constant and to show what effect changes to tax rates in isolation would have had on the post-tax income distribution. We call this the 'what if?' approach. It yields the apparently quite different conclusion that cuts in tax rates over the 1980s made a significant contribution to the rise in overall income inequality.

As with so much of the analysis in this area, it is not possible to say that one of these answers is right and the other wrong. They are simply answers to different questions. The Jenkins study (based on the 'actual payments' approach) answers the question 'how much redistribution is the tax system doing now compared with a decade ago?'. The answer to this is that it is doing about as much in the late 1980s as in the late 1970s. This is because the pre-tax distribution has become much more unequal, and so even with lower tax rates, the tax system is still doing plenty of redistribution. The Johnson and Webb study (based on the 'what if?' approach) is answering the question 'what would the post-tax distribution look like if the tax system had not been changed?', and the answer to this is 'much less unequal'.

In this chapter, we use the 'actual payments' approach. That is to say, we examine the actual amounts of tax paid and benefits received by each type of household in our data and examine how those have changed over time. The alternative would have been to have attempted to apply the tax and benefit system of the early 1960s to the pre-tax income distribution of the 1990s to see how different the post-tax distribution would have looked.

There are two main reasons why we do not adopt this 'what if?' strategy. In the first place, one of the assumptions that we make when we impose an old tax system on a current distribution is that there would be no changes in people's behaviour if the old system were to be reintroduced. This assumption may be just about tenable for minor changes from one year to the next, but is a scarcely credible assumption for the effects of a complete reversal of three decades of tax and benefit changes. The second reason for looking simply at actual payments of tax (and receipt of benefit) is the huge practical difficulties of attempting to simulate all the rules and regulations of the tax and benefit system as it prevailed in the early 1960s.

THE EFFECTS OF THE CURRENT TAX AND BENEFIT SYSTEM

In this section, we look at the direct taxes people pay and the social security benefits that people receive. It should also be stressed that our analysis is partial in that we restrict ourselves to looking at the effects of taxes on income and of cash transfers. For the whole analysis of the book is based on an income measure that includes cash benefits and is net of direct taxes. We look neither at indirect taxes and taxes on capital, nor at the distributional effects of spending on, for example, health and education. The interested reader is referred to the regular articles in the CSO's *Economic Trends* publications for details of the effects of carrying out this sort of exercise in one particular way.

Income Tax

We begin our analysis with income tax, which is the most important source of government income, accounting for around a quarter of all tax revenues in 1995. To examine how the income tax system impinges on households at different income levels, we start by ranking the population by household equivalent disposable income (as before).[1] Next, we examine the amount of income tax paid by each household as a share of total household pre-tax income. Finally, we report the twenty-fifth percentile, the fiftieth percentile, and the seventy-fifth percentile of these average tax rates for each decile group. This approach has the advantage of providing some indication of the number of taxpayers in each decile group as well as the tax burden amongst those who are paying tax. It is also less sensitive to extreme values which can distort the average figure for a particular decile group.

---

[1] We do not in this case make an adjustment to the incomes of the richest households (because of problems in adjusting the FES tax payments data on a consistent basis) but this does not affect the ranking of individuals.

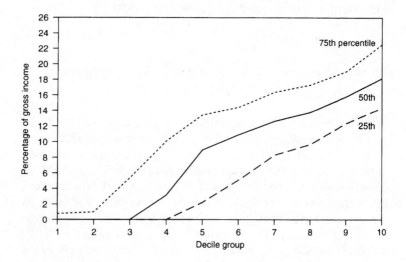

Fig. 7.1. Income tax as a percentage of gross income by decile group, 1991–93

The results of this analysis for income tax in 1991–93 are shown in Fig. 7.1.

If we consider first the middle line of the three in Fig. 7.1, this shows the average tax rate of the person whose household tax burden is at the mid-point of the tax burdens of those in each decile. For shorthand, we shall refer to this person's experience as 'typical' for the decile. Figure 7.1 indicates that the poorest 30 per cent of the population typically pay no income tax at all, whilst the typical income tax burden for the richest tenth of the population is around 18 per cent.

This latter figure prompts the question 'how can an average income tax rate of 18 per cent be typical for the richest tenth in the population, when the standard rate of income tax in 1991–93 was 25 per cent and the higher rate 40 per cent?'. The answer to this lies in the structure of tax allowances and bands. Every taxpayer has an annual tax-free personal allowance, and any income up to this amount is completely free of tax. Additional tax allowances are available to those who are married, those who have mortgages,[2] and those who make pension contributions, amongst others. For this reason, although the marginal tax rate on the last pound may be 25 per cent, because of the various allowances and exemptions, the average tax rate will often be well below this. Furthermore, although there is a 40 per cent rate of income tax, this only applies at relatively high incomes, so that only around 7 per cent of tax-payers, and an even smaller proportion of the whole population, face a 40 per cent rate on their final slice of income.

Comparing the middle (median) line in Fig. 7.1 with the twenty-fifth and seventy-fifth percentiles illustrates the point that, even for those in the same income decile, there can be a significant variation in the burden of income tax. In part, this is because certain forms of income are not subject to income tax, and so the *composition* of total income will affect the tax bill, and not just the total amount. Also, this variation arises because the income tax system takes no account of the number of children in a household, whereas our decile ranking treats children as reducing a household's standard of living. Consequently, two households in the same decile may have quite different incomes before equivalisation, and hence different tax bills, even though their underlying 'living standards' (on our definition) may be the same. The income tax system is not a tax on equivalent income.

Other important reasons for the difference in income tax treatment of people with similar incomes arise from the particular way in which married couples are taxed. Compare, for example, two married couples, one with one earner on £40000 gross and the other partner not working, the other with each partner earning £20000. The gross income is the same in each case. But the net income of the two-earner couple will be rather higher than that of the one-earner couple for they will each have a full tax-free allowance, each will benefit from the 20 per cent band, and neither will be even close to paying any tax at 40 per cent. The single earner will have only one allowance and, assuming no additional tax allowances, will be paying a 40 per cent marginal rate on the last £10000+ of earnings. In Fig. 7.1, the single-earner couple will appear in a lower net income decile and will be paying a higher proportion of their income in tax.

[2] Since 1983, under the Mortgage Interest Relief At Source (MIRAS) system, this relief has been administered via a reduction in mortgage payments rather than a reduction in income tax. For the purposes of this analysis, however, it is counted as a tax reduction.

These features of the tax system have the result that people who are, from an income point of view, 'equals' are often treated unequally. In addition, the income tax results in a degree of re-ranking. Households that were better off than others before tax can end up worse off after tax, and vice versa.[3] To some extent, this illustrates a general feature of what Fig. 7.1 shows. Because deciles are ranked by net income (that is, after tax), those with very high tax payments will have been pushed down the distribution. The lines are likely to be flatter than would have been the case if deciles had been ranked by gross income.

It is also apparent from Fig. 7.1 that at least a quarter of the individuals in the bottom decile group are paying some income tax. In general, these will be poorer individuals who have some interest income from a bank or building society account that is automatically taxed at source, even though their total income is well below the tax threshold. It is possible for poorer households to apply for refunds of this tax, but many have failed to do so where they continue to hold their money in accounts where tax is deducted at source. This is likely, in part, to reflect the rather small amounts of money at stake. Other very low income groups paying some income tax will include those with a cash income over £3500 per year but a very small equivalent income because of their family size.

Right at the top of the distribution, everybody is paying some tax, with a typical average tax rate of around 18 per cent and a quarter having average tax rates of over 22 per cent. The difference between such people in the middle and even three-quarters of the way up the top decile and those right at the top is again worth stressing. Somebody earning, say, £100000 per year would have an average tax rate of 35 per cent or so, but such people are on incomes well in excess of those displayed here.

The main point to note is that income tax is a *progressive* tax. That is, on the whole, the average rate increases with income right across the income distribution. This progressivity derives very largely from the existence of the tax-free allowances at the bottom of the income tax schedule. The fact that the vast majority of taxpayers pay income tax at the same *marginal* rate—the basic rate—does *not* prevent the system from being progressive.

[3] See, for example, Aronson, Johnson, and Lambert (1994) for an empirical analysis of this issue.

Box 7.1. The income tax system

Income tax is paid on taxable income. Taxable income is that income above tax-free allowances. Each individual has a tax-free allowance of £3765 in 1996–97.

On the first tranche of income above the allowance—the first £3900 in 1996–97—a rate of 20 per cent is charged. On the next £21 600, the basic rate of 24 per cent is charged. Any income in excess of this attracts a marginal tax rate of 40 per cent.

Tax liability is reduced by tax reliefs—for example, on contributions to pension schemes, and mortgage interest relief—and for married couples by the Married Couple's Allowance.

Because of the tax-free allowance, the *average* rate of income tax is always less than the *marginal* rate.

National Insurance Contributions

Figure 7.2 shows the incidence of employee National Insurance contributions (NICs) on the same basis as Fig. 7.1.

The general pattern of Fig. 7.2 is that average employee NIC rates rise with income, but fall back for the highest decile group. The NIC system as it affects lower and middle earners is broadly progressive, as those below a lower earnings limit pay no NICs, and those above pay only a 2 per cent rate on this first slice of earnings. Above this point, there is a 10 per cent charge on extra earnings. The 'regressive' section at the top of the income distribution (where average NIC rates fall though income rises) is brought about by the ceiling on employee NICs. Above this level, no further NICs are payable and so at higher incomes the NIC bill represents a declining proportion of total income. It is also the case that the self-employed, who pay a relatively low average rate of NICs, are over-represented in the top decile group and, of course, unearned incomes, such as income from investments or private pensions, of which there are some very high examples, are not subject to NICs.

It is also noticeable from Fig. 7.2 that, even in the fifth decile, more than a quarter of people pay no NICs as employees or through self-employment. This is, in part, because pensioners are exempt from

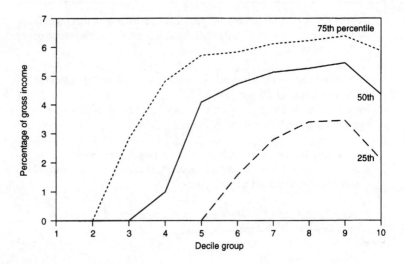

Fig. 7.2. NICs as a percentage of gross income by decile group, 1991–93

NICs, and also because most forms of unearned income such as investment income, pensions, and benefits are not subject to NICs. As a result, it is possible to have an income close to the national average but to be paying no NICs.

It should be noted that Fig. 7.2, along with the rest of our analysis, does not take any account of the impact of employ*er* NICs. This is a regrettable omission, but unfortunately the Family Expenditure Survey data on which our results are based do not indicate the level of NICs paid by employers in respect of employees in the survey. In principle, we would wish to include the effects of employer NICs. Since there is no upper limit on the amount of NICs that employers have to pay, to include employer NICs would make the overall NIC system appear much less regressive.

Social Security Benefits

Turning now to the benefits system, we consider separately the role of means-tested and of National Insurance (NI) benefits. Means-tested benefits are those payable only to claimants with income and assets

Box 7.2. National Insurance contributions for employees

NICs are payable only on earned income (or profits for the self-employed). For employees, no NICs are payable until earnings reach the *Lower Earnings Limit* (LEL)—£61 per week in 1996–97.

Once earnings exceed this level, a charge of 2 per cent of the LEL is payable. On every pound above the LEL, NICs are charged at a rate of 10 per cent.

This rate continues until the *Upper Earnings Limit* (UEL) is reached—£455 per week in 1996–97. No more contributions are payable on earnings above this level.

below certain specified levels. The main means-tested benefits are Income Support (cost £17.2 billion in 1995–96) for those on low incomes and not in full-time employment, Housing Benefit (cost £10.5 billion) for those whose rent is high relative to their income, and local tax rebates (Council Tax Benefit since 1993, Community Charge Benefit prior to that, and Rate Rebates when the old rates system was still in operation). In addition, Family Credit is available to low-income families in work with children.

National Insurance benefits are available to those who have made the requisite number of NI contributions and who satisfy a contingency such as old age, unemployment, sickness, maternity, or widowhood. The main NI benefits in terms of expenditure are the state Retirement Pension (involving expenditure of £30 billion in 1995–96) and Incapacity Benefit which replaced the old Invalidity Pension in 1995. Other NI benefits include Widow's Pension and part of the Job Seeker's Allowance (the replacement for Unemployment Benefit).

These two categories of benefit account for more than three-quarters of total benefit spending and include all those benefits whose role is either income maintenance or earnings replacement. There is a final and smaller class of benefits which we do not consider here which is the group of *contingent* benefits that are designed to assist with particular costs. Rights to them depend neither on income nor on the NI contribution record, but on certain contingencies such as having children or being disabled. The most important is Child Benefit (cost £6.5 billion in 1995–96) which is payable in respect of any children, whilst others include Disability Living Allowance.

Figure 7.3 shows what proportion of the disposable household income of individuals in different decile groups is made up of means-tested benefits, and Fig. 7.4 repeats the analysis for National Insurance benefits.

Figure 7.3 shows very clearly the way in which means-tested benefits such as Income Support and Housing Benefit are concentrated on poorer households. For those most heavily dependent on means-tested benefits, they can easily form three-quarters or more of total household income. Even in the third decile group, more than a quarter of people are obtaining a third of their income from such benefits. It is also clear that not all people who fall in the bottom decile group are in receipt of means-tested benefits. The full-time self-employed with low profits (or losses) are one group who may have very low incomes but not be eligible for means-tested assistance; others might include pensioners with very low housing costs and little private income. The upward slope of the fiftieth percentile line between the first and second deciles is again a result of the fact that the deciles are ranked by net income (that is, income including benefits); high levels of

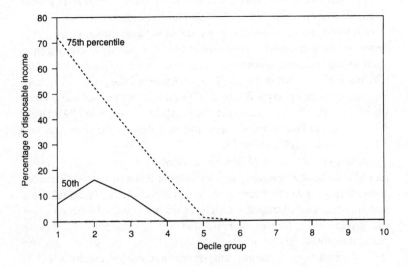

Fig. 7.3. Means-tested benefits as a percentage of disposable income by decile group, 1991–93

means-tested benefit, especially Housing Benefit, might well be adequate to shift people out of the bottom decile.

Figure 7.4 shows the share of NI benefits in income declining less rapidly as income rises. They play an important part for some households even in the fifth and sixth deciles. One of the key determinants of the contribution of NI benefits is the incidence of pensioners at different points in the income distribution, since the Retirement Pension is by far the largest NI benefit. In particular, there is a heavy concentration of pensioners in the second decile group, which explains why the median share of NI benefits actually rises very substantially between the first and second deciles.

It is worth stressing that the figure does not indicate that there is no receipt of NI benefits right at the top of the income distribution, just that fewer than 25 per cent of those in these deciles receive any such benefit. As we saw in Chapter 2, there are a number of pensioners right at the top of the income distribution and they too will be receiving income from the basic Retirement Pension. Indeed, richer pensioners can receive rather more of this than poorer pensioners

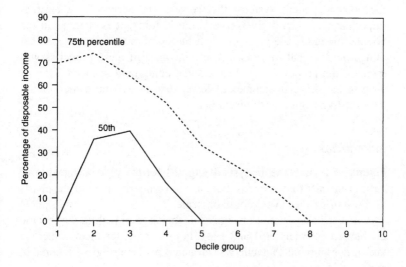

Fig. 7.4. National Insurance benefits as a percentage of disposable income by decile group, 1991–93

because it is more likely that both partners in a couple will have an adequate NI record to provide full entitlement to a basic pension in their own right.

National Insurance benefits (paid for out of today's NI contributions) clearly do favour poorer households but not nearly as strongly as means-tested benefits which are heavily concentrated in the poorest decile groups. Whereas there is still significant receipt of some NI benefits in the upper half of the distribution, entitlement to means-tested benefits has been largely exhausted by the time the fifth decile is reached.

TRENDS 1961–93

Figures 7.1 to 7.4 provided a snapshot of the way in which the direct tax and benefit system affects the distribution of income. The next stage is to investigate how the incidence of taxes and benefits between households at different income levels has changed.[4] In considering what follows, it is important to bear in mind that changes in incidence will result from two different causes. The first is, of course, changes in the tax and benefit systems themselves. The second is at least as important and should always be borne in mind. It is changes in the population and in the level and distribution of pre-tax incomes. Other things being equal, increases in the incomes of the top deciles will increase the proportion of their income being paid in income tax. A drop in the original incomes of lower deciles will raise the amount they receive from means-tested benefits.

Income Tax

Figure 7.5 repeats the analysis of Fig. 7.1 for the years at the start of each of the last four decades. For ease of comparison, all of the charts in the set of four have a common scale.

A first striking feature of Fig. 7.5 is the large growth in the income tax burden between 1961 and 1981. For example, the median income tax burden in the fifth decile was around 4 per cent of gross income in

---

[4] For detailed analyses of how changes to the tax and benefit system have affected people's incomes over various periods between 1979 and 1995, see Johnson and Stark (1989), Johnson and Webb (1993), and Giles and Johnson (1994).

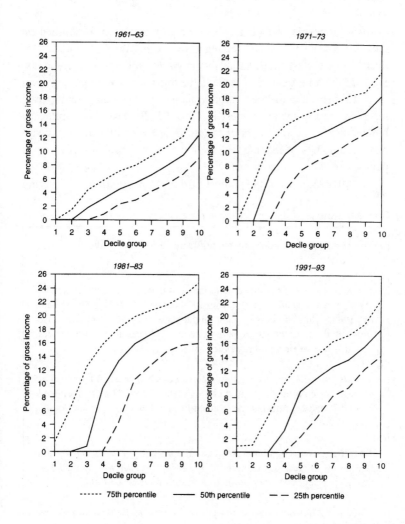

Fig. 7.5. Income tax as a percentage of gross income by decile group, over four decades

1961–63, around 11 per cent in 1971–73, and almost 14 per cent in 1981–83. It is also interesting to note that although this figure fell back over the following decade, it still stood at more than 8 per cent in 1991–93.

Also noticeable from Fig. 7.5 is the way in which the burden on those on lower incomes has changed over time. In 1961–63, most of those in the lowest decile groups were paying little or no income tax. By 1971–73, one-quarter of those in the third decile were paying more than 12 per cent of their gross income in tax—an average rate reserved mainly for the richest in 1961–63. By 1981–83, whilst the highest taxpaying members of the third decile continued to pay significant amounts of income tax, the lowest-taxed half of this group is paying virtually no tax at all.

By 1991–93, most of the poorest fifth of the population were paying

Box 7.3. Some of the main income tax changes

The *basic rate* on earned income fell from 32 per cent in the late 1960s to 25 per cent in the early 1990s.

The *highest rate* of tax on earned income was 96.25 per cent in the late 1960s. It was cut substantially from 1979 onwards, reaching 40 per cent by 1988. The numbers paying higher-rate tax did, however, rise substantially over the 1980s, partly due to cuts in certain tax allowances and also due to rises in pre-tax incomes.

The *tax-free allowance* for a single person reached nearly 30 per cent of average male earnings in the mid-1960s and early 1970s. It fluctuated around 20 per cent during the 1980s, hitting a low of 16 per cent in the early 1990s.

Tax relief on mortgage interest was available at the taxpayer's marginal rate on all interest payments until 1974. By 1995, it was available at just 15 per cent and only on the interest on the first £30000.

Child Tax Allowances were phased out in the late 1970s and replaced by Child Benefit.

Independent taxation of husband and wife replaced the old system of joint taxation in 1990.

virtually no tax, whilst the burden on those on slightly higher incomes had also fallen rather significantly. This reflects a combination of cuts in tax rates and changes in the pre-tax distribution. At the bottom end of the distribution, there has been a growth in the number of unemployed people and lone parents, two groups who typically have very low tax liabilities. At the top, pre-tax incomes have grown very dramatically, which would tend to increase average tax liabilities. However, the large cuts in the standard and higher rates of income tax have tended to offset this change.

National Insurance Contributions

Results on a similar basis for employee and self-employed National Insurance contributions are shown in Fig. 7.6.

A first point to note about Fig. 7.6 is that the left-hand scale runs only up to 7 per cent, which indicates that, compared with income tax, NICs on employees and the self-employed have historically accounted for a much smaller proportion of household gross incomes.

One clear pattern from Fig. 7.6 is that, over time, the impact of NICs has become considerably more progressive. In 1961, NICs were a mixture of flat-rate and earnings-related contributions, which meant that they bore relatively heavily on low earners. In the mid-1970s, they were reformed to be proportional to earnings, with the rate of employee NICs being raised several times in the early 1980s. In 1985 and again in 1989, the structure was further reformed, with the aim of reducing the burden of NICs on the lower-paid.

These reforms are reflected in Fig. 7.6, which shows that in 1961–63 and 1971–73, NICs bore most heavily on those in the third or fourth income deciles, whereas in 1981–83, the proportionate burden was actually heaviest in the eighth decile. In 1991–93, the burden was still heaviest near the top of the income distribution, whilst the burden on those with the lowest incomes had fallen considerably.

The size of the NIC burden clearly increased at least until the early 1980s. Indeed, the need to raise extra revenue from NICs was largely responsible for the move away from a flat-rate charge towards more of an earnings-related system. There were increases in the rate of NI at the start of the 1980s, but the restructuring at the end of that decade in which the allowance-like structure was introduced at the bottom involved a substantial reduction in the overall NI burden. The increase in the main rate from 9 per cent to 10 per cent in 1994 is not captured in the years analysed here.

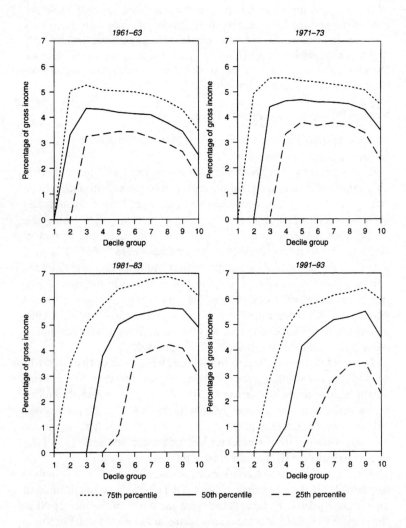

Fig. 7.6. NICs as a percentage of gross income by decile group, over four decades

Box 7.4.   Changes to employee National Insurance contributions

Until 1975, there was a substantial flat-rate element to NICs. On top of that, there was a graduated payment which rose from 4.25 per cent to 5 per cent of earnings (up to the UEL) over the period.

From 1975 until 1985, employees simply paid a proportion of all earnings (including earnings below the LEL) once they earned above the Lower Earnings Limit. The rate rose from 5.5 per cent in 1975 to 9 per cent by 1985.

In 1985, a series of lower NI rates were introduced for lower earners, with the main one remaining at 9 per cent.

Since 1989, reaching the LEL has only incurred a charge of 2 per cent. The main rate of 9 per cent (10 per cent since 1994) was then charged only on earnings between the LEL and UEL.

Changes over time have clearly made the NI system substantially more progressive than was the case back in the 1960s and early 1970s. The virtual phasing-out of the flat-rate element was largely responsible for this. With the exception of the Upper Earnings Limit, it has (for earners) come to look increasingly like the income tax system. The abolition of the UEL would largely complete this process and get rid of the downturn in average burdens currently observed in the top decile. Losing the 2 per cent charge on incomes below the Lower Earnings Limit would clearly reduce the sharpness of the rise at lower incomes.

Whilst employer contributions are not included in these charts, it should also be noted that in 1985 a major change took place in the structure of employer NICs. This was the abolition of the Upper Earnings Limit for employers, which meant a large increase in liability in respect of high earners. The revenue from this change was used to finance cuts in the burden on low-paid employees, and this switch greatly increased the progressivity of the overall NIC system.

Means-Tested Benefits

Figure 7.7 shows the results for means-tested benefits, expressed in this case as a proportion of household incomes *after* tax and NICs have been deducted. It indicates that there have been dramatic changes over the last three decades in the contribution of means-tested

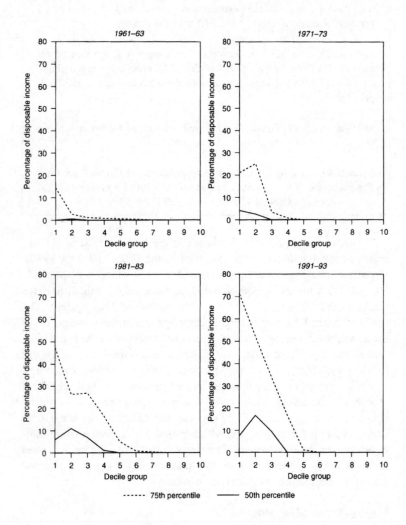

Fig. 7.7. Means-tested benefits as a percentage of disposable income by decile group, over four decades

benefits to household incomes. In 1961, the main means-tested social security benefit was National Assistance, the forerunner of Supplementary Benefit and then Income Support. This was originally intended to be a residual benefit for those whose National Insurance benefits were for some reason inadequate to bring them up to a minimum standard of living. In fact, National Assistance always had a greater role than had been envisaged by Beveridge when designing the post-war social insurance system, but even so, the amounts paid were relatively small, as the top-left panel of Fig. 7.7 indicates.

By 1971–73, means-tested benefits were starting to become more important, with a national system of Rent Rebates and Rates Rebates newly in place.[5] Furthermore, a new in-work means-tested benefit called Family Income Supplement (FIS) had just been introduced to top up the earnings of low-paid workers with children. FIS was initially a very small-scale benefit but, as with housing and local tax benefits, its importance steadily grew over the coming decades.

The growth in the role of means-tested benefits can be seen very clearly in the pictures for 1981–83 and 1991–93. In 1981–83, those in the bottom decile most heavily dependent on means-tested benefits were receiving almost a half of their income from this source, and this figure had risen to more than 70 per cent by 1991–93.

Five key factors explain the growth in the role of means-tested benefits in household incomes, particularly by the late 1970s and 1980s. These are

1. *A large growth in the client groups most likely to be on low incomes.* The main groups involved are lone parents, the unemployed, and older pensioners.
2. *A tightening of the rules for eligibility to NI benefits.* Not only did the relative level of certain NI benefits drop, but growing numbers were deemed ineligible because of an inadequate record of NI contributions.[6] Groups hit particularly badly included the young unemployed, others with variable work histories, and the self-employed.

---

[5] In the 1960s, such rebates as were available were often administered locally and took the form of a reduced rent. Unfortunately, our data for the 1960s do not enable us to identify cases where the rents or rates charged have been subsidised in this way.

[6] See Atkinson and Micklewright (1989).

3. *A relative fall in the generosity of NI benefits for those receiving them*. A number of changes, including a switch from earnings indexation to price indexation of the basic pension and abolition of earnings-related supplements to short-term NI benefits, meant that growing numbers needed means-tested top-ups to their NI benefits.

4. *An increase in the generosity of means-tested benefits*. A clear (though relatively modest) example of this is FIS (which later became Family Credit), whose role as a top-up to low pay has expanded considerably. More importantly, for much of the period, the value of the safety-net Supplementary Benefit / Income Support line has risen in real terms, thereby increasing claimant numbers. For example, research for the Social Security Committee (1993) suggests that the number of pensioners receiving Income Support in 1989 would have been 50 per cent lower if the real level of benefit were to be returned to its 1979 level.

5. *Rising rents*. Housing Benefit has become increasingly important as rents have risen in both the private and public sectors despite falls in the numbers of individuals in these sectors. Among social tenants (those housed by local authorities and Housing Associations), the increased receipt of Housing Benefit is associated with reduced capital subsidies which have pushed up rents and so pushed up Housing Benefit payments. This change can be seen as a move from a universal benefit in the form of low rents towards a more means-tested system in the form of benefit payments to those on low incomes.

Taken together, these factors help to explain why, for the poorer groups, means-tested benefits have in many cases become the main source of household income.

National Insurance Benefits

Many of these trends are also apparent in Fig. 7.8, which shows the results for National Insurance benefits. It shows that for many of the poorest households, NI benefits have always been a very important source of income. In 1961–63, half of the poorest decile group obtained more than 70 per cent of their income from this one source, with a similar pattern holding for 1971–73. Over the following two decades, however, a growing proportion of the poorest groups have

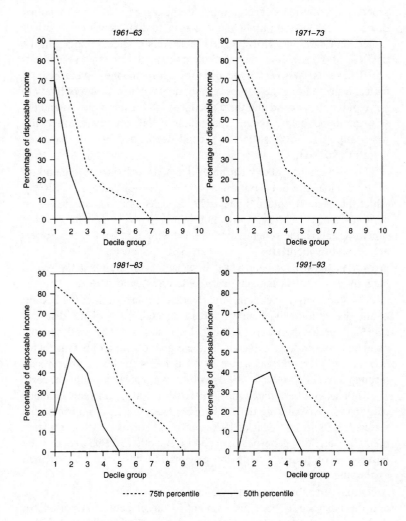

Fig. 7.8. National Insurance benefits as a percentage of disposable income by decile group, over four decades

been receiving little or no help from NI benefits.[7] By the end of the period, half the poorest decile had no such benefits at all.

In part, this declining role for NI benefits at the very bottom of the income distribution reflects some of the aspects of benefit policy discussed earlier, including the decision to link pensions to prices rather than earnings. However, the shifts also reflect changes in the position of the pensioner population, who are now the main recipients of NI benefits. Whereas three decades ago pensioners were heavily concentrated among the poorest group, they are now far more likely to be found in the second and third deciles, with the unemployed and the self-employed below them. Consequently, NI benefits are actually more important to those in the second decile in 1991–93 than they were in 1961–63.

This pattern has been reinforced by a growth in entitlements to benefits under the State Earnings-Related Pension Scheme (SERPS), which has been in place since 1977. SERPS pensions are based on earnings since 1977 for those not in company or private pension schemes, and as the scheme matures, growing numbers of pensioners are receiving increasing amounts of SERPS pensions. The planned growth in SERPS pensions has, however, been significantly reined back by government legislation in 1988 and again in 1995.

In tandem with a reducing importance for those right at the bottom of the distribution, NI benefits spread through more of the distribution. The period between 1971 and 1981 was one where NI benefits extended their influence even up to the eighth decile group. One factor behind this increase was a rise in NI benefits to the long-term sick through Invalidity Benefit (IVB), which was introduced in the early 1970s. The numbers receiving IVB rose steadily, and then dramatically particularly during the 1980s. However, this in turn prompted a series of cuts in the level of IVB, and presaged its eventual replacement by the less generous Incapacity Benefit in 1995. Again, the movement of pensioners up the distribution has been important in explaining this trend.

National Insurance benefits, which had been designed by Beveridge and the post-war Labour government to target people in periods of their greatest need—when their incomes from work were lost because of unemployment, sickness, widowhood, or old age—became progressively less well targeted on the poorest over the 1970s and

---

[7] See Webb (1994) for a comprehensive analysis of the changing role of NI benefits.

1980s. This reflected increased benefits which drove some recipients into higher deciles, the increased private incomes of some groups, notably pensioners, with rights to NI benefits, and increased incomes from other family members. This last effect would include the impact on household income of having a spouse in work whilst claiming Invalidity Benefit or Unemployment Benefit, for example.

To some extent, NI benefits, which are being gradually cut back, are a victim of the change in the underlying distribution of income. There being more people in the relevant groups has resulted in their cost being raised. More better-off people in these groups has meant that the spending on them has become less easy to justify in terms of poverty alleviation. It is becoming less easy to identify in advance who will be poor in the future, and it is on being able to identify such contingencies that the NI system is based. Things have changed since the system was introduced some fifty years ago. There are new contingencies such as lone parenthood not catered for by NI benefits, and old contingencies such as unemployment have become too widespread. Old age, which was always seen as one of the best predictors of poverty, no longer is, and much pension spending goes to those with incomes outside the bottom reaches of the income distribution.

CONCLUSIONS

Over the last three decades, there have been major changes to the direct tax and benefit system. The overall burden of income tax has risen over the period as a whole, despite the cuts of the 1980s, but the structure remains highly progressive. National Insurance contributions began the period as a relatively regressive tax, but successive structural reforms have removed most of the regressive elements. Means-tested benefits began the period as a relatively modest part of the incomes even of the poorest households, but have grown dramatically in importance for many poorer households, particularly in the 1980s. National Insurance benefits remain an important part of the benefit system, particularly for pensioners, although they are now less central to the incomes of the very poorest households and are no longer the exclusive province of the poor.

8 Poverty

INTRODUCTION

Our concern so far has primarily been with the current pattern of UK income inequality and with how that pattern has changed over the last three decades. One of the reasons for a concern with inequality is to assess how far there are groups in society whose living standards may differ significantly from those of the majority. In particular, the presence of individuals or households whose living standards are markedly *below* those of the majority is likely to be of particular concern. If these individuals have an unacceptably low standard of living, then improving that living standard is likely to be a policy priority. For this reason, we focus in this chapter on the issue of what can broadly be called 'poverty'. First, we consider how poverty might be defined and measured, and then we provide some indication both of the total numbers and of the sorts of individuals who might be said to be living in poverty in the UK on various definitions.

WHAT DO WE MEAN BY POVERTY?

In order to quantify the extent of poverty in a given country at a point in time, it is necessary to take two decisions. First, what should be the yardstick by which poverty should be measured? For example, is it about incomes or expenditure? Or perhaps about patterns of social participation? Second, having determined the yardstick, at what point on the yardstick is the 'poverty line' to be set? For example, in the case of incomes, should it be incomes below some fraction of the national average? Or incomes close to minimum social security levels?

A wide variety of responses to these questions have been advocated in the academic literature on poverty measurement, and a range of methods have been implemented by official bodies in various countries in an attempt to put these measures into practice. Since our concern in this book is primarily an empirical one—in this case, to assess how many people are poor, what sort of people are poor, and how these patterns have changed—we do not present here an extensive theoretical discussion of alternative approaches to measuring poverty.

However, since it is important to understand the nature of the poverty measures that are used in the remainder of this chapter, we do present here a brief summary of some of the main alternative approaches that are available. This summary is based largely on Nolan and Whelan (1996) to which the interested reader is referred for a comprehensive discussion of the issues surrounding the measurement of poverty and deprivation.

Nolan and Whelan identify six different approaches to measuring poverty. These are as follows:

1. *Budget standard approach.* Since food is a fundamental necessity, a traditional starting-point for measuring poverty has been to work out how much money is needed to purchase a 'nutritionally adequate' diet. Since there are other necessities, some adjustment is then made to this figure to allow for non-food expenditures. This adjustment may take the form of identifying and costing a list of other necessities or simply scaling up the food cost estimate by some standard multiplier. Poverty is then defined as having an income insufficient to cover the cost of a nutritionally adequate diet together with other necessities of life. The budget standard approach has a long historical pedigree in the UK, with the pioneering studies of Rowntree (1901) being based on this method.

2. *Food ratio method.* A related approach is founded upon the observation by Engel (1895) that the share of total income spent on necessities tends to fall as income rises. In the light of this relationship, one possible poverty yardstick is the proportion of income spent on food (or necessities more generally). Where necessities account for a large part of total household spending, a household may be regarded as being in poverty. One difference between this method and the budget standard approach is that no attempt is made in this case to assess what nutritional adequacy involves, or to identify a list of other 'necessary' items. Food ratios are used in Canada as the basis for the 'Low Income Cut-Offs' presented in official statistics.

3. *Social security levels.* In this case, the poverty yardstick is simply income and the poverty line is drawn relative to the prevailing rates

of social security benefits in the country concerned. To the extent that minimum social security rates reflect society's views about the amount of money needed to attain a minimum acceptable standard of living, this cut-off point has some attractions. This method was pioneered by Abel-Smith and Townsend (1965) and underlies the 'Low Income Families' statistics that are presented later in this chapter.

4. *Consensual income poverty lines.* If poverty is related to society's views about an acceptable standard of living, then one approach to determining an income poverty line is to assess popular views on this issue on the basis of large-scale surveys. A variety of methods have been adopted, including asking respondents to specify the incomes that hypothetical families would need to reach a certain standard of living, and asking individuals about their feelings about their own standard of living. The answers to these questions are then used to link welfare levels with incomes. Finally, a 'critical' welfare level is selected and mapped onto a corresponding income level, and that income level is then used as the poverty line. An example of the application of this technique is van Praag, Hagen-aars, and Van Weeren (1982).

5. *Purely relative poverty lines.* A widely used method of defining poverty is to relate incomes (or expenditures) to some proportion of the prevailing national average. National average can be defined as mean or median, the proportion used can vary, typically ranging from 40 to 60 per cent, but the general principle is that poverty is to be defined wholly by distance from national average income. Statistics of this sort are now widely used in comparative studies (for example, O'Higgins and Jenkins (1990) on the members of the European Community). As we have already seen, this approach also forms the basis for the principal UK series, 'Households Below Average Income', although the various thresholds used are not explicitly identified as 'poverty lines'.

6. *Style of living / deprivation.* In this case, the poverty yardstick is not simply income but a wider measure of living standard as indicated by individual lifestyles and ability to participate in society. The approach taken by Mack and Lansley (1985) and described later in this chapter falls within this broad category. In brief, it first involves identifying 'socially defined necessities' (which may include ownership of consumer durables or ability to participate in a hobby or leisure activity) on the basis of a survey of

the population. On the basis of this survey, poverty is then defined in terms of the number of people lacking a significant number of these 'necessities'.

Each of these approaches has its own strengths and limitations. Three points are, however, of particular importance when using methods such as these to measure poverty. The first is that each is to some extent arbitrary, and none can claim to be an 'objective' measure of poverty. For example, it might be assumed that the inability to purchase a basket of basic necessities would meet anyone's definition of poverty. However, it would not be possible to construct a basket of 'necessities' that attracted universal assent. Views about the composition of the basket would vary from individual to individual, and no single basket could be regarded as objectively 'correct'. Similarly, if the lack of 'socially defined necessities' were agreed to correspond to notions of poverty, the decision to regard the lack of three necessities (rather than two or four) as poverty would inevitably be arbitrary.

A second point to note is that each measure of poverty is to some extent 'relative' in that it is likely to change as the living standard of society as a whole changes. This is most obviously true with the purely relative poverty lines such as half national average income, but is also true even of the budget standard approaches. For example, whereas a television might have been regarded as a luxury item half a century ago, it would be quite reasonable today to regard a family that could not afford a television as being 'deprived'. In other words, it is not possible to define poverty, and in particular trends in poverty, in any meaningful way without some reference to the prevailing living standards in society as a whole.

Third, poverty is likely to be multifaceted, and hence, as with inequality, a single method of measuring poverty is likely to give only a partial picture. For example, low income may coexist with a relatively high level of welfare, whilst those with relatively high incomes may none the less experience a variety of forms of deprivation. Indeed, it is the contention of Nolan and Whelan (1996) that the poor as defined by income and the poor as defined by 'style of living' measures may in many cases be quite different groups, and that a proper understanding of poverty needs to take account of a variety of perspectives rather than a single summary measure.

The main purpose of this discussion has been to highlight the difficulties of arriving at a single measure of poverty, and also to indicate the range of alternative approaches that are available. In the remainder of this chapter, we present results on levels of, and trends in, UK poverty, based on a variety of definitions. The approaches have been selected on the basis of three objectives:

(1) to provide estimates of numbers in poverty for a relatively recent period;
(2) to offer some time-series dimension on how the numbers affected have changed over a period of years; and
(3) to be nationally representative and hence based on a significant sample of the UK population.

There are only two main approaches that have been adopted in the UK that satisfy these objectives, and each represents a different method of measuring poverty according to the Nolan and Whelan typology. The two methods are *Households Below Average Income* ('purely relative') and *Low Income Families* ('social security levels'). A more limited coverage is also provided by the two *Breadline Britain* studies (based on the 'style of living/deprivation' approach), and some results from those studies are also presented below.

HOW MANY PEOPLE ARE POOR?

As one would expect from the discussion so far, there is no single answer to the question of how many people in the UK are poor in the early 1990s. In this section, we provide a range of estimates that vary according to the poverty line and method of measuring living standards that are used. We report results from two main sources: the 'Households Below Average Income' series, whose definitions we have used throughout this book; and the 'Low Income Families' series, first published by the (then) DHSS and now continued by the Institute for Fiscal Studies. The main features of each approach are summarised in Table 8.1.

As we have seen, these are by no means the only ways of looking at poverty, and we also include some useful complementary data from other sources. These include figures on durable ownership, numbers below proportions of average spending, and, in particular, an approach from the 'Breadline Britain' studies based on access to essentials.

Table 8.1. Key features of alternative studies of poverty

	Households Below Average Income	Low Income Families
Period covered	DSS: 1979 to 1992–93 IFS: 1961 to 1991	DSS: 1972 to 1985 IFS: 1979 to 1992
Main data source	Family Expenditure Survey	DSS administrative data and Family Expenditure Survey
Sample size	7000 households each year	7000 households each year and 1 per cent sample of benefit recipients
Poverty line	Fractions of mean income	Basic benefit levels
Measure of living standards	Income	Income
Unit of assessment	Household	Benefit unit

The Households Below Average Income Approach: Levels and Trends

The Households Below Average Income (HBAI) approach uses the Family Expenditure Survey to make estimates of the number of people in the country with household incomes below various fractions of the national mean. The results for 1992–93 are summarised in Table 8.2.

The yardstick of 50 per cent of national average income is one commonly used in cross-national studies of poverty (see, for example,

Table 8.2. Key HBAI results for 1992–93

	Individuals with household income below various percentages of the mean (millions)		
	40%	50%	60%
Income before housing costs	5.4	11.5	17.5
Income after housing costs	8.0	14.1	18.8

Source: DSS, 1995.

EC (1991)). On this basis, and using household incomes before housing costs have been met, 11.5 million people, or around one in five of
the UK population, are living in poverty. On the alternative measure
of living standards, which takes account of variations in people's
housing costs, the number in poverty would be 14.1 million. Table
8.2 shows both the importance of the particular definition of income
that is used and the impact of the precise poverty line chosen. A
poverty line set at 40 per cent of average income would, in the case of
income before housing costs, reduce measured poverty by half compared with the 50 per cent threshold.

The advantage of using a range of lines as set out in Table 8.2 is
that the poverty measure obtained from a single line may be highly
sensitive to the precise positioning of that line. This point is illustrated
in Fig. 8.1, which shows, for 1992–93 and for equivalent income
before housing costs, the distribution of income for each of our six
broad family types together with a vertical line indicating half
national average income.

For some family types, such as working-age couples with children,
the precise location of the poverty line will have relatively little effect
on the numbers within the group appearing 'in poverty'. This is

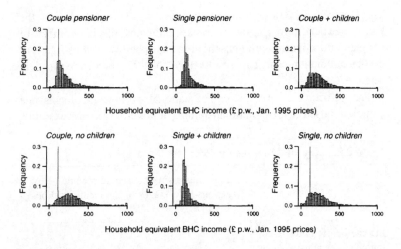

Fig. 8.1. Distribution of income and 'poverty line' set at half average
income by family type, 1992–93

because the incomes of this group are relatively evenly spread and no particular low-income threshold has any special significance for them. However, as Fig. 8.1 shows, this is certainly not the case for groups such as lone parents and pensioners, whose incomes are highly concentrated around benefit levels which are in turn quite close to half national average income. It is clear that the use of a slightly lower poverty line would take millions of pensioners and lone parents out of measured poverty, whereas a slightly higher line would bring millions in. This is largely what is responsible for the patterns shown in Table 8.2, which indicates that there are many individuals with incomes between 40 and 50 per cent of the national average.

One problem with this approach to measuring poverty is that it is highly dependent on the level of mean income, which in turn may be skewed upwards by some very high incomes at the top of the income distribution. It could be argued that developments at the top of the distribution should not affect the measurement of poverty.

A measure less sensitive to such effects would be one based on a proportion of the *median*—that is, the point in the middle of the distribution—so that we would be comparing the lowest incomes with the middle. On this basis, we would find 8.8 million individuals with incomes below half the AHC median and 6.3 million with incomes below half the BHC median. At 40 per cent of the median, these numbers fall to 4.7 million and 2.7 million respectively.

Changes over Time

In some ways, it is easier to interpret changes over time in some of these numbers than to be clear of the implications of having a certain number of people below a line at any given point in time. In Figs. 8.2(a) and 8.2(b), we show the proportion of the population below 40 per cent, 50 per cent, and 60 per cent of national average income on the BHC and AHC income measures respectively in each of the last thirty-three years.

What is striking about Fig. 8.2 is that, whether the definition of income is before or after housing costs, and whether the poverty line is set at 40 per cent, 50 per cent, or 60 per cent of national average income, the numbers below the line have risen dramatically since the late 1970s. Using the 50 per cent threshold, the proportions have risen from 6 per cent to 20 per cent of the population (BHC) or from 7 per cent to 25 per cent (AHC) since 1977. These charts do, however,

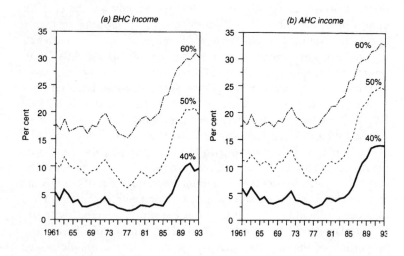

Fig. 8.2. Percentage of the population below 40 per cent, 50 per cent, and 60 per cent of national average income

demonstrate that the choice of poverty line can still have an important bearing on the precise description of trends as well as levels. If the 40 per cent BHC threshold is used, the poverty rate could be said to have risen more than fivefold, whereas at the 60 per cent threshold, the rate has barely doubled.

It is also apparent from Fig. 8.2 that the fluctuations in the numbers in poverty on these definitions were much less pronounced over the 1960s and early 1970s than in the 1980s, with the BHC poverty rate (at the 50 per cent threshold) remaining within the 8–12 per cent band for the whole period 1961–74.

It should be stressed that the various thresholds used in Fig. 8.2 have all risen much faster than prices over the past three decades, and particularly so over the 1980s. A strictly 'absolute' approach to poverty would imply that all these trends are measuring is inequality rather than poverty, since the poverty yardstick is constantly being increased. We consider this point further below.

The Low Income Families Approach: Levels and Trends

The Low Income Families (LIF) series has two main distinctive features compared with the HBAI series. First, the 'poverty line' used is based on Income Support benefit levels rather than being some fraction of national average income. Second, individuals' living standards are assessed in terms of the family (or, more precisely, 'benefit unit') of which they are a member, rather than of the whole household. This is a direct consequence of using a benefit level as the point of comparison, since benefit entitlement is calculated at this level of aggregation. This definition means that each unmarried adult or couple within the household (together with dependent children if present) is treated as a separate unit. The income of each benefit unit is assumed to be independent of and unaffected by the incomes of any other benefit unit in the household.

Apart from these differences, the other main difference between HBAI and LIF is that in the latter, income (or 'net resources') is always expressed after the deduction of housing costs such as rents or mortgage interest.

Table 8.3 shows, for 1992, the number of individuals and benefit units with net resources at different levels based on the LIF methodology. Note that the groups are cumulative, so that everyone who appears in the first row of the table will also be counted in the second and third rows.

Table 8.3 indicates that in 1992 there were 2.9 million families who were not in receipt of Income Support (IS) but who had net resources below IS levels. This category includes those who are entitled to Income Support and who fail to take up their entitlement for some reason, and those who are not entitled but who none the less have low

Table 8.3. LIF results for 1992

(Millions)	Families	Individuals
Not receiving Income Support, with resources below IS levels	2.9	4.7
Receiving Income Support, or with resources below IS levels	7.9	13.6
Receiving Income Support, or with resources up to 140 per cent of IS levels	10.7	18.6

Source: Social Security Committee, 1995.

measured resources. Non-entitlement can arise from, for example, ownership of assets in excess of £8000 or working more than 16 hours per week. The 2.9 million families contain 4.7 million individuals (including children).

If poverty is defined as being 'on or below' IS levels, then the numbers involved rise to 7.9 million families containing 13.6 million individuals. A much higher poverty line of 140 per cent of IS levels (sometimes described as being 'on the margins of poverty') embraces 10.7 million families or 18.6 million individuals.

Comparing the results for HBAI (after housing costs) and LIF for this period, one sees 14.1 million individuals with incomes below half the average and 13.6 million on or below IS levels. This might appear to indicate a broad measure of consensus on the basis of different definitions, though, of course, many of those defined as poor on one of these measures will not appear to be so on the other, and vice versa. Furthermore, it should be stressed again that both are highly sensitive to small changes in the precise threshold chosen.

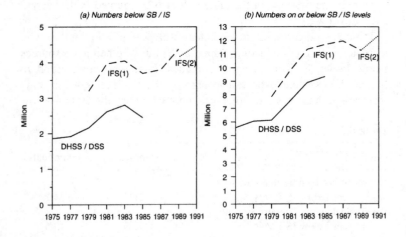

Fig. 8.3. Low Income Families

Changes over Time

One of the main difficulties of using the LIF series over time is in constructing a consistent picture over the many changes in definition that have affected the series since its inception. Figure 8.3(a) provides the available information for the numbers strictly below Supplementary Benefit/Income Support levels, whilst Fig. 8.3(b) shows numbers 'on or below' SB/IS levels. In each case, there are three data sources: DHSS/DSS based on Department of (Health and) Social Security publications, IFS(1) constructed for Social Security Committee (1992), and IFS(2) constructed for Social Security Committee (1995).

Considering first Fig. 8.3(a), which deals with the numbers not receiving SB/IS but with resources below SB/IS levels, it is clear that there is a marked discontinuity between the old DHSS/DSS series and the new IFS series. The old series is included to provide some information on trends in the 1970s which is not available on the other (we believe better) measure.[1]

On the revised definition, Fig. 8.3(a) shows a modest rise in the numbers below SB/IS levels from around 3 million in 1979 to more than 4 million in 1991. In part, this reflects the fact that SB/IS levels themselves rose in real terms over the period, particularly for older pensioners, and this could in itself increase the numbers with relatively low incomes. It is sometimes argued that this apparently perverse result—that raising benefit levels could increase poverty—wholly invalidates the LIF approach. However, if a poverty line is chosen specifically because it bears some relationship to the general prosperity of the country rather than simply being frozen in real terms, then the fact that more people are below that line remains a significant finding.

A more dramatic pattern can be seen in Fig. 8.3(b), which includes the individuals below SB/IS levels together with those receiving SB/IS. Here, the rise is much greater, with the numbers affected rising from just under 8 million in 1979 (on the IFS series) to more than 12 million in 1991. A comparison of Figs. 8.3(a) and 8.3(b) indicates that

---

[1] Part of the reason for this is a set of changes in the definitions and coverage of the statistics made for the 1979–89 edition, notably an extension to cover Northern Ireland (rather than just Great Britain) and a switch from 'normal' to 'current' income. This latter change ended the practice of treating those who had been unemployed for less than 13 weeks as if they were still employed at their previous wage. The revised definition is thought to provide a better guide to current living standards.

most of this increase is coming from those who are actually receiving SB/IS. The reasons for this increase are discussed more fully below, but the main factors are rises in unemployment and lone parenthood.

Comparing the trends in HBAI and LIF results, it is apparent that whilst both indicate a significant rise in poverty over the course of the 1980s, the two series differ in the timing and extent of the rise. Whereas the HBAI figures do not take off until the second half of the 1980s, there is a marked increase in the LIF figures during the recession of the early 1980s. Also, when the HBAI numbers do increase, the rate of increase is much more rapid than that reflected in the LIF series. These differences reflect the fact that HBAI numbers are most sensitive to changes in national average income (which rose rapidly in the late 1980s) whereas LIF reflects more closely the numbers dependent on benefits (which tend to rise particularly rapidly during times of recession).

One can reasonably argue that it is a serious weakness of an approach to poverty measurement based on numbers below a proportion of the average income that such measures behave in this way. One would expect periods of recession, not periods of boom, to be associated with increased poverty. The fact that the non-employed fell further behind the employed during the late 1980s is not necessarily a reason for believing that levels of poverty would have increased. The increase in the numbers out of work during recessions might, however, reasonably be expected to be associated with a rise in poverty.

Other Indicative Measures

The HBAI and LIF statistics are based purely on information on incomes. As we have seen, a more complete picture of poverty can be obtained by using a broader range of measures. In this section, we consider patterns of durable ownership, levels of household expenditure, and the more comprehensive 'style of living' studies undertaken for the 'Breadline Britain' programmes.

Durable Ownership

The most comprehensive relevant figures relate to durable ownership among those in the bottom income groups. Table 8.4 (based on FES data as with the above analyses) shows the proportion of the bottom decile (before housing costs) with access to various durables in

Table 8.4. Access to consumer durables of bottom decile group

Percentage of individuals in household with access to a:	1962–63	1972–73	1982–83	1992–93
Telephone	8%	20%	58%	78%
Washing machine	—	54%	79%	89%
Fridge or fridge-freezer	—	52%	95%	99%
Car	—	26%	44%	56%
Video cassette recorder	—	—	—	68%
Central heating	—	20%	46%	73%

selected years since the early 1960s. Note that information about the ownership of most durables shown is not available until the early 1970s.

These figures provide a good idea of the sorts of durables to which those who might be counted as poor had access. In 1992–93, a majority had access to each of the durables specified. Over the 1970s and 1980s, the proportion of individuals in this lowest income group with access to all the consumer durables rose very substantially. There were especially big increases in the proportions with a telephone and with central heating. So, despite no increase in the real incomes of this group, one can argue that in some senses at least they did become better off.

Expenditure

Another purely relative poverty line could be derived by following the HBAI approach, but using expenditure as the chosen measure of the standard of living rather than income. As we saw in Chapter 4, there are considerable advantages to using spending to measure living standards, particularly as a means of capturing the longer-term aspects of households' well-being.

The number of people with spending below half the average (mean) expenditure in 1992–93 was slightly *less* than the number whose incomes were below half the mean. There were also fewer people whose expenditure was below 40 per cent and 60 per cent of the mean, compared with the income measure. This is shown in Table 8.5, which sets out the numbers of people living in households with expenditure below 40 per cent, 50 per cent, and 60 per cent of mean expenditure in

Table 8.5. HBAI: income and expenditure compared, 1992–93

| | Individuals with household income/ expenditure below various percentages of the mean (millions) | | |
	40%	50%	60%
Income before housing costs	5.4	11.5	17.5
Expenditure	4.8	10.6	16.8

1992–93, comparing these with the numbers with income below these same proportions of mean income. What this table shows is that the incidence of poverty is somewhat lower if spending rather than income is used as a yardstick.

The incidence of poverty is also on a rather different set of people under the expenditure measure. We return to this in the next section, which examines what sort of people are poor.

Turning to the changes over time, we find that over the whole of the 1970s and up until the late 1980s, there were actually more people living below the spending-based poverty lines than below income-based ones. Figure 8.4 compares the proportion of the population below half the mean income over time with the proportion below half the mean spending. (The patterns revealed are similar to those for below 40 per cent and 60 per cent, which are not illustrated here.) We can see that up until the early 1970s, the numbers in poverty were very similar under both measures. Over the middle part of the 1970s, the proportion below half average income fell back quite sharply, as incomes across all but the very bottom of the income distribution fell. By contrast, the proportion below half average spending remained much more steady, at around 10 per cent, falling by around only 1 percentage point to about 9 per cent by 1977.

Both the income and spending measures showed a similar rate of growth over the recession in the early part of the 1980s and then fell back again. It was only over the mid-1980s that the numbers in poverty under the income measure grew much faster, so that the numbers in each group were roughly similar by the late 1980s.

Figure 8.4 also shows clearly the seemingly perverse effect that some recessions have had on purely relative poverty measures such as the HBAI ones. In the recessions of the early 1970s and of the early 1990s, the poverty headcount actually fell, whether it is income or

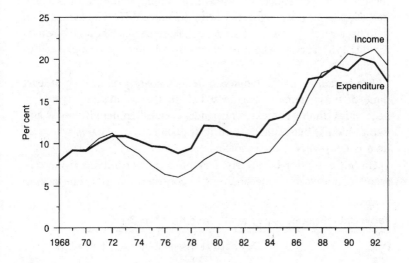

Fig. 8.4. Percentage of population below 50 per cent of mean income and 50 per cent of mean expenditure compared

spending that is used in the assessment. This shows that the middle and top of the distributions must have been worse hit by the recession than the bottom. Similarly, in periods of rapid, but uneven, economic growth, large gains for those at the top along with smaller gains for others can result in a very rapid increase in the numbers in poverty.

Breadline Britain

The method adopted for these studies was to interview a sample of around 1800 individuals to obtain information about the items that respondents thought to be necessities and also about the items that the respondents themselves lacked. The items that were put to respondents included: ownership of durables such as a fridge, television, or washing machine; housing quality indicators such as absence of damp or presence of an indoor toilet; dietary options such as three meals a day for children and a roast joint once a week; and items connected with 'social participation' such as a 'best outfit' for special occasions, the ability to have children's friends round for tea periodically, and money for Christmas presents for family and friends. Table 8.6 shows, for 1990, the percentage of respondents saying that each of selected items was a 'necessity' together with the percentage who lacked each item because they could not afford it. The table also shows the results for the 1983 survey where a shorter list of items was suggested to participants.

The Breadline Britain approach defines poverty as lacking at least three of those items that at least half of the population regards as necessities. In this context, a respondent 'lacks' an item if they would like to have it but cannot afford it. It does not cover cases where an item is not owned out of choice.

On this definition of poverty, lacking an indoor toilet, a damp-free home, and a washing machine would constitute poverty, but lacking

Table 8.6. Selected survey results from Breadline Britain

Item	Described as essential by (%)		Lacked by (%)	
	1990	1983	1990	1983
Damp-free home	98	96	2	7
Indoor toilet (not shared)	97	96	*	2
Heating for living areas	97	97	3	5
Bed for everyone	95	94	1	1
Decent state of decoration	92	—	15	—
Fridge	92	77	1	2
Insurance	88	—	10	—
Washing machine	73	67	4	6
Regular savings of £10 per month	68	—	30	—
Hobby or leisure activity	67	64	7	7
A television	58	51	1	*
A 'best outfit' for special occasions	54	48	8	10
A night out fortnightly	42	36	14	17
A car	26	22	18	22
Restaurant meal monthly	17	—	22	—
A home computer	5	—	16	—

* = Less than 0.5%.
— = Not asked in 1983.
Source: Frayman *et al.*, 1991.

only two of these plus a car would not constitute poverty. Both the 'three items' threshold and the '50 per cent' threshold are inevitably arbitrary. The former threshold is chosen on the grounds that the correlation between low income and lack of 'necessities' becomes markedly stronger at the 'three or more' threshold. This reasoning is, however, circular, since if the point of the exercise was partly to get away from measuring poverty simply in terms of low income, then it seems odd to define a poverty line at least in part with reference to the incomes of the sample.

It is instructive, none the less, to examine the numbers below a variety of thresholds, and results from the Breadline Britain study are presented in Table 8.7. Again, this shows results for the two years in which the survey was conducted—1983 and 1990.

Table 8.7 shows that in 1990 around 11 million individuals were in households that lacked three or more of the 'necessities' identified by respondents to the survey. Of these, more than half lacked five or more necessities and a third lacked seven or more. These all mark an increase by comparison with the earlier survey in 1983.

Over a period during which average incomes rose sharply, benefits rose more or less in line with retail prices, and unemployment fell, the results in Table 8.7 prompt the question 'how can the number of people lacking necessities have risen?'. The basic answer is that the number of things defined as necessities rose. The key differences between the results for 1983 and those for 1990 (as shown in detail in Table 8.6) are set out below:

1. Of the items included in both surveys, there was a general increase in the proportion of the population who felt them to be necessities.
2. The proportion of people lacking any one of these items had almost without exception stayed the same or fallen. Thus in 1990, more

Table 8.7. Breadline Britain: numbers lacking 'necessities'

Number of 'necessities' lacked	Number affected (millions)	
	1983	1990
Three or more	7.5	11
Five or more	5.5	6
Seven or more	2.8	3.5

Source: As Table 8.6.

people had damp-free homes, more people had inside toilets, more
people had heating, and so on.

3. The reason why the second Breadline Britain study appears to find
   more people 'in poverty' than the first one is simply that the list of
   'necessities' has become longer. This in turn is the result of two
   effects. The first is a rise in expectations, where some items previously below the 50 per cent threshold are now deemed to be
   'necessities' by a majority of the population (such as a 'best outfit'
   for special occasions). The second, and more important, effect is
   that new items have been added to the list—of which decoration,
   insurance, and regular savings are the most significant. Each of
   these is deemed a necessity by a majority of the population and is
   also lacked by relatively large numbers.

These features suggest that it would be misleading to compare the
Breadline Britain results for 1983 and 1990 and to infer that there had
been a 'rise in poverty' of 3.5 million. Had the list of items from 1983
simply been repeated in 1990, and even allowing for the change in
public views about necessities, it seems highly unlikely that the rise in
poverty would have been anything like as large (or indeed that the
numbers involved would even necessarily have risen). On the other
hand, if the underlying methodology is accepted, it could be argued
that the 1990 list is more satisfactory since it includes a number of
items omitted in 1983 that the public regards as necessities and that
significant numbers of people lack.

WHO ARE THE POOR?

So far, this chapter has looked just at the numbers of individuals who
might be considered poor on a number of definitions. As with previous chapters, on the income distribution, we now go on to look at
the family type and economic status composition of those in the
poorest groups.

HBAI

We again start by looking at those below fractions of the national
average. On this occasion, we restrict ourselves to looking at the
composition of those with below *half* average income, and in the

years 1979 and 1992–93. The changes between these dates are the most interesting in terms of composition, and these are also years that allow a direct comparison with results from figures based on numbers at and below Income Support levels.

On the AHC measure, we find in Table 8.8 that over half of all those living in single-parent families had incomes below half the average in 1992–93. The next most over-represented group were the single pensioners, over a third of whom were in this position. None of the relativities between these groups is surprising, given what we already know. But if we change our poverty line to 40 per cent of the average (again using the AHC measure), pensioners stop being over-represented. Around 10 per cent of both couple and single pensioners are then found below the poverty line compared with 14 per cent of the population as a whole. Virtually 30 per cent of single parents are found even in this lower income category. As was illustrated in Fig. 8.1, exactly where the poverty line is drawn makes a big difference to the composition of 'poverty'.

Over the period since 1979, the biggest change was the major deterioration in the position of lone parents relative to other population groups and in particular relative to pensioners. But no group escaped the general increase in the numbers below half the average. Back in 1979, there were some groups who were virtually certain of not being in this income bracket, particularly working-age individuals without children. This changed fundamentally over the period.

When the population is divided by economic status, virtually none of the families containing just full-time workers are below even 50 per

Table 8.8. Percentages of family types with incomes below half the contemporary mean

| | Before housing costs | | After housing costs | |
	1979	1992–93	1979	1992–93
Pensioner couple	16	25	21	26
Single pensioner	16	25	12	36
Couple with children	7	20	8	24
Couple without children	4	10	5	12
Single with children	16	43	19	58
Single without children	6	18	7	22
All family types	8	20	9	25

cent of the mean in either period. Among the group made up of one full-time worker and one non-worker, however, there has been a major increase in the incidence of poverty on either baseline. For example, on the AHC basis, just 4 per cent of this group had incomes below half the average in 1979 as against 16 per cent in 1992–93. At this date, three-quarters of the unemployed had incomes below half the average and 62 per cent had incomes below 40 per cent.

Taking the HBAI approach but using spending rather than income as the living standards measure leads to many more pensioners and fewer families of working age being classified as poor. This is because, as we saw in Chapter 4, pensioners' spending tends to be low compared with the average, whilst there are a considerable number of low-income non-pensioner families whose spending is relatively high.

LIF

As with the HBAI series, LIF results provide information on the family type and economic status of individuals in the various low-income categories. Table 8.9 provides a family-type breakdown of the LIF results for 1992 at various income thresholds.

Considering first the group not receiving Income Support but with resources below IS levels, three main causes can be identified:

1. *Non-take-up of benefit entitlement*, particularly in the case of pensioners. Many pensioners with state pensions and small amounts of other income may have a small entitlement to Income Support each

Table 8.9. LIF results for 1992 by family type

(Number of individuals, millions)	Not receiving IS, resources under 100% of IS level	Receiving IS, or resources under 100% of IS level	Receiving IS, or resources under 140% of IS level
Total	4.7	13.7	18.6
Of which:			
Married pensioners	0.5	0.8	2.0
Single pensioners	0.8	2.0	3.0
Couple with children	1.5	3.7	5.2
Couple, no children	0.6	1.1	1.7
Single with children	0.2	3.0	3.1
Single, no children	1.2	3.1	3.6

Source: Social Security Committee, 1995.

week. This may go unclaimed due to a variety of factors such as
the effort involved in claiming, the possible 'stigma' associated
with receipt of welfare benefits, lack of knowledge of the complex-
ities of the system, or possibly just delays in receipt after a claim.[2]
Whatever the reason, such individuals are likely to find themselves
with resources slightly below IS levels.

2. *Support from other household members*, particularly among young
single people. The unit of assessment used in LIF is that of the
family or 'benefit unit', which means that a single adult (for
example) still living at home with parents will be treated as a
separate unit. In some cases, such individuals will have little or
no independent income (and hence appear in the LIF tables as
being well below IS levels) but may choose to be supported by
parents or other household members rather than claim benefit.

3. *Disqualification from benefit.* Some groups such as those in full-
time work may have low incomes but not be entitled to Income
Support, since it is available only where no family member works
16 hours per week or more. Two groups are particularly likely to
appear in the LIF figures for this reason—the full-time self-
employed with low reported profits and others in full-time work
who have large mortgages. For the latter group, even if their
disposable income after meeting the mortgage takes them below
IS levels, no social security help is available because they are
working full-time.

A rather different set of circumstances affects the 9 million individuals who are in families receiving Income Support. Amongst pensioners, as Table 8.9 indicates, it is single pensioners who are the
main recipients. This group is dominated by older, female pensioners
who will often be dependent on relatively small amounts of state and
private pensions, in many cases paid to them on the basis of a late
husband's employment or National Insurance contributions.

Amongst those of working age, the main reasons for being dependent on Income Support are unemployment, lone parenthood, and
sickness or disability. In 1992, claimant unemployment averaged
around 2.8 million, and the majority of these individuals (together
with their families) would have been receiving Income Support.
National Insurance Unemployment Benefit was available to those

[2] See DSS (1995) for estimates of the extent of non-take-up of income-related
benefits, and Fry and Stark (1990) for a discussion of the determinants of non-take-up.

unemployed people who had an adequate record of NI contributions and had been unemployed for less than a year, but only around one-third of the unemployed fell into this category. Furthermore, for those with children, the level of Unemployment Benefit plus Child Benefit was insufficient to lift them above IS levels and so means-tested Income Support was payable in addition.

Lone parents, who numbered around 1.4 million in the UK in 1992, are another group with a high incidence of Income Support receipt. For lone parents with young children in particular, it can be difficult to combine childcare with paid employment, and around three-quarters of lone parents have insufficient earnings or other income to lift them clear of the Income Support system.

The long-term sick and disabled form a final major group of Income Support recipients that has also been growing rapidly in recent years. The reasons for this growth are not well understood, given the general improvement in health standards, but it seems likely that in some cases this is a disguised form of unemployment or early retirement, where employers or GPs are willing to indicate that individuals are sick in order to help them receive more favourable treatment from the benefit system.[3]

Table 8.9 also identifies a further total of around 5 million individuals who are not receiving Income Support but who have incomes in the range 100–140 per cent of IS levels. The circumstances of this group are fairly diverse, but more than 2 million are pensioners who have sufficient pensions and private income to lift them above basic IS levels, but who may still be receiving some means-tested benefits such as Rent Rebates or local tax rebates. Amongst those of working age, the main factors will be low-paid employment, high housing costs, or receipt of sickness-related benefits that are above IS levels.

The message of LIF therefore is that poverty tends to be associated with unemployment, lone parenthood, sickness or disability, and, to some extent, old age.

Breadline Britain

This pattern is also reflected in the 1990 Breadline Britain results, though these provide far less detail on the composition of the poor.

---

[3] The related phenomenon of a growth in the number of recipients of Invalidity Benefit is analysed in Disney and Webb (1991).

What they do show is that two-thirds of lone parents, and around half of the unemployed, lack three or more 'necessities'. The rate of poverty amongst those in full-time work is found to be low in the 1990 Breadline Britain study. However, because of the large size of the total group, full-time workers still make up one-third of those who lack three or more necessities.

Somewhat surprisingly, pensioners appeared to have a lower incidence of poverty than might have been expected from other sources. Frayman *et al.* (1991) suggest that this may reflect the fact that pensioners 'expect less out of life', and hence are less likely to indicate that they would actually like one or more of the items that they lack.

HOW HAS THE COMPOSITION OF THE POOR CHANGED?

In terms of *trends* in the composition of the poor, the HBAI and LIF series provide the best information. As noted above, there are problems of comparability between the two Breadline Britain studies, although they do appear to indicate that lone parents and the unemployed were the groups most at risk of poverty in both 1983 and 1990.

Turning then to the LIF results over time, Table 8.10 shows the numbers of individuals on or below SB/IS levels in 1979 and in 1992. It shows a rise of around 6 million in the number of individuals in families on or below basic benefit levels. What is immediately apparent

Table 8.10. LIF results for 1979 and 1992 by family type

(Number of individuals, millions)	Receiving SB/IS, or resources under 100% of SB/IS level	
	1979	1992
Total	7.7	13.7
Of which:		
Married pensioners	1.2	0.8
Single pensioners	2.2	2.0
Couple with children	1.6	3.7
Couple, no children	0.4	1.1
Single with children	1.1	3.0
Single, no children	1.1	3.1

is that this rise has occurred despite an improvement in the position of pensioners. The number of pensioners at these relatively low incomes has actually fallen from around 3.4 million in 1979 to 2.8 million in 1992, and would have fallen substantially further had it not been for real increases in the value of the IS line for this group. The main reasons for the improved position of this group have been a rise in pension income, whether from the state earnings-related scheme (SERPS) or from a former employer, and a rise in income from investments.

Amongst non-pensioners, the two groups with the largest proportionate rises are lone parents and the single childless. As regards lone parents, not only has the total number of lone parents in the population risen markedly over the period, but the proportion dependent on Supplementary Benefit/Income Support has also risen. One of the main reasons for this latter trend is that the composition of the lone-parent population has changed, with a smaller proportion of divorced, separated, or widowed mothers and a growing proportion of never-married lone mothers with younger children. Opportunities for well-paid employment may be very limited for such mothers, partly because of childcare responsibilities as noted earlier and also because many have relatively low levels of qualifications and hence low earnings potential.

The rise in the number of single childless people on low incomes is mainly a reflection of the rise in unemployment, but this group also includes growing numbers of full-time students with relatively low incomes. The growing number of couples on or below IS levels is also mainly an unemployment effect, but the related causes of involuntary 'early retirement' and growing incidence of long-term sickness are also contributory factors.

SUMMARY AND CONCLUSIONS

We have examined trends in UK poverty using two main statistical approaches: the HBAI series, which compares household incomes with fractions of the national average, and the LIF series, which compares family incomes with prevailing benefit levels. These have been supplemented by data on ownership of durables, data on patterns of household expenditure, and the results of the Breadline Britain surveys, which examine popular views about 'social necessities' and the extent to which such necessities are lacked by different groups.

Each approach is based on different data sources, definitions, and time periods. In terms of the numbers in poverty in the early 1990s on the various definitions, we found nearly 14 million individuals in families that were receiving Income Support or had incomes below the IS line. A very similar number had household incomes below half the mean (on the after-housing-costs measure), but this number is very sensitive to the cut-off point chosen. Eight million had incomes below 40 per cent of the mean on this measure, 5.4 million on the before-housing-costs measure.

As regards trends in poverty over time, the long-run perspective available from the HBAI approach indicates that from the early 1960s to the mid-1970s, the poverty rate was relatively stable. The late 1970s and 1980s were clearly very different, with the HBAI poverty rate climbing rapidly in the later part of the period whilst the LIF-based figure rose most significantly during the recession of the early 1980s.

Whilst the total numbers in poverty on these various definitions have risen markedly, the composition of the poor has changed significantly. Both the LIF and HBAI studies indicate that at the start of the 1990s, pensioners are far less likely to be found amongst the poorest groups than at the end of the 1970s, whereas there is a growing although still small incidence of very low incomes amongst those in some form of employment. Throughout the period, the unemployed and lone parents have been at considerable risk of poverty, and the size of both groups has grown markedly.

9   Income Dynamics

INTRODUCTION

The analysis of the preceding chapters has been based on a snapshot, or in some cases a series of snapshots, of the UK income distribution. This has enabled us to address questions such as 'at a given point in time, how many people are poor?' and 'what sort of people are poor?'. In this chapter, we move beyond this kind of snapshot analysis and look at some of the dynamic properties of the income distribution, largely by tracing individuals' incomes and circumstances over a period of time. This allows us to provide some answers to questions such as 'for how long are people poor?' and 'by what processes do people enter / leave poverty?'.

Unfortunately, the main data source that we have used so far—the Family Expenditure Survey—does not allow us to provide ready answers to this sort of question because it is a 'cross-section' survey. In a given year, a representative sample of the population is contacted and interviewed, and the following year a different cross-section of the population is interviewed. What the FES does not do is to track the same people over time. As a result, we have no way of knowing, for example, whether the households that were rich or poor in a given year of FES data would still have been rich or poor one year later.

The lack of UK data that track the same individuals regularly and over a prolonged period (known as 'panel' data) has long hampered those interested in how people's position in the income distribution changes over time. This situation is, however, beginning to change, and a panel study begun in the early 1990s—the British Household Panel Survey—is starting to provide some answers to these questions.

We begin the chapter, however, by giving a flavour of the sorts of results that can be derived from panel data that have been running for many years, such as those that have been collected in the US. Next, we give some preliminary results from the British Household Panel Survey and provide a brief summary of some other examples of UK panel data. In the UK context, this is still a rather early stage in the development of this type of dynamic work and results are still based only on quite short periods of data. This is one area in which we can expect our understanding to be considerably enhanced in the relatively near future.

Finally, we examine what can be done to look at trends in individual incomes not simply from one year to the next but over a complete lifetime. In order to do this, we use successive cross-sections of FES data to create synthetic or 'pseudo-cohorts' of individuals whose living standards can be traced over a period of decades. This approach, which is explained more fully below, provides some insights into how incomes vary over an individual's lifetime and also how the pattern of lifetime incomes may itself be undergoing change.

US PANEL DATA

One of the longest-established panel studies is the US Panel Study of Income Dynamics (PSID), which started in 1968 and is still continuing. Because the study has been in existence for such a long period, it can be used to examine changes in living standards over significant sections of the lives of panel members.

One example of use of the PSID that illustrates the attractions of panel data is that by Walker *et al.* (reported in Walker with Ashworth (1994)), which examines patterns of childhood poverty in the US. The sample chosen was all children in the panel who were born in the years 1968 to 1972 inclusive, which produced a sample of around 1300 children. The living standard of each child was measured in terms of its family income relative to a 'low-cost' budget, set roughly 25 per cent above the official US poverty line. Children with family incomes below this level were deemed to be 'in poverty'.

In analysing the children's experience of poverty over the course of their 15-year childhood, the authors of the study identify six 'types' or patterns of poverty which are listed in Table 9.1.

One interesting feature of this typology is that a child who is observed in a single cross-section to be 'in poverty' could in fact fall into any one of the six categories. In other words, this could be a transient experience, never again repeated throughout childhood, it could be a recurrent (albeit short-term) experience, or it could be a permanent feature of life. Only with the benefit of panel data is it

Table 9.1. 'Types' of poverty

(a)	Transient	A single spell of poverty lasting a single year.
(b)	Occasional	More than one spell (during childhood) but none lasting more than one year.
(c)	Recurrent	Repeated spells of poverty, some separated by more than a year but some exceeding a year in length.
(d)	Persistent	A single spell of poverty lasting between two and thirteen years.
(e)	Chronic	Repeated spells of poverty never separated by more than a year of relative prosperity.
(f)	Permanent	Poverty lasting continuously for fifteen years.

Source: Walker with Ashworth, 1994, p. 122.

possible to make these distinctions and to examine which of these experiences is the most common.

Walker *et al.* find that 62 per cent of the children in their sample experience no poverty on this definition at any point in their childhood. Of the remainder, Fig. 9.1 shows the breakdown between the various categories of poverty.

Figure 9.1 shows that more than 40 per cent of the children who are ever poor experience recurrent poverty, albeit temporarily alleviated from time to time, whereas more than a quarter have only a relatively fleeting experience of poverty lasting for at most a year. At the other extreme, one in twenty of the children who are ever poor are continuously poor for the whole of their childhood.

There are considerable potential benefits for policymakers, amongst others, of looking at the dynamics of the income distribution. Suppose, for example, that it was possible, on the basis of panel data, to identify in advance the types of children most at risk of spending their entire childhood in poverty. In this case, policy could be devoted to early intervention that might prevent the experience of poverty from becoming permanently established. This might be a much more effective form of policy activity than a generalised anti-poverty strategy that gave equal weight to the one in four children whose experience of poverty was brief and not repeated.

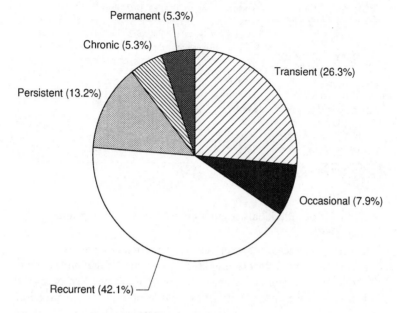

Fig. 9.1. Patterns of childhood poverty in the PSID
Source: Walker with Ashworth, 1994, p. 123.

UK PANEL DATA

The British Household Panel Survey

The first year's fieldwork for the British Household Panel Survey (BHPS) was undertaken in 1991. About 7500 households from around Great Britain were contacted and, of these, roughly 5500 agreed to take part in the survey. Adult members of each household were asked a series of questions about their incomes, employment, family circumstances, and attitudes on a range of issues. Questions were asked not only about their current circumstances but also about their position in each of the preceding 12 months. A year later, the same individuals were interviewed again (together with any new members of their households), and this process is to be continued for as long as the BHPS remains in existence. Table 9.2 summarises

Table 9.2.   Individual response rates to Waves 1, 2, and 3

	Wave 1 respondents	New respondents	Wave 2 respondents	New respondents	Wave 3 respondents
Wave 1	**9912**				
Died/Out of scope	−139				
Eligible at Wave 2	9773				
Wave 2	8567	+892	**9459**		
Died/Out of scope	−121		−336		
Eligible at Wave 3	8446		9123		
Wave 3	7617		8205	+812	**9017**

Source: *BHPS News*, Issue no. 8, Autumn 1994

the number of individuals responding to each of the first three waves of the BHPS.[1]

Table 9.2 indicates that in the 5500 participating households in Wave 1, 9912 adults were prepared to respond to the survey. Over the following 12 months, a small number (139) either died or moved beyond the scope of the study (for example, abroad), but in principle the remaining 9773 could have been interviewed again. Of these people, around 1200 either refused to participate a second time or could not be contacted. This would have left a sample of 8567 in Wave 2 but for the fact that a number of additional adults became eligible to be surveyed. These included the 15-year-old children from Wave 1 who reached 16 and thus entered the sample, and also some adults who entered the household of original participants (through marriage, for example). These 'new entrants' topped up the original sample to give a total of 9459 adults interviewed in Wave 2.

By a similar process, the numbers interviewed in Wave 3 (1993) fell further to 9017, of whom 7617 had also been interviewed in both Waves 1 and 2. The way in which some individuals drop out of panel data of this sort is known as 'attrition'. Where attrition is rapid or where the people who drop out are very different in some respect from the people who continue to participate, this can limit the usefulness of panel data. Statistical techniques that can attempt to overcome biases of this sort do, however, exist, and reweighting of data is one example.

[1] A more detailed description of the BHPS, together with a range of analyses of the first two waves, is contained in Buck *et al.* (1994).

One application of panel data is to examine what happened from one year to the next to the living standard of different groups. Most of the analysis that follows concentrates on assessing the persistence of low incomes, but such data can equally well be used to examine the position of those on higher incomes. However, preliminary analysis suggests that it is amongst the lowest income groups that income mobility is most marked, and therefore where panel data can provide most insights.

Table 9.3 presents the results of one such analysis on the basis of the first two waves of the BHPS.

Each cell of Table 9.3 provides information on what happened to the real disposable household income between Wave 1 (1991) and Wave 2 (1992) of a different group of the population. The first row relates to those who were in the poorest tenth of the population in Wave 1, the second row to those who were not in the poorest tenth in Wave 1, and the final row to the whole Wave 1 sample. Similarly, the first column of results is for those who were in the poorest tenth in Wave 2, the second column for those not in the poorest tenth in Wave 2, and the final column for the whole Wave 2 population.

Considering first the bottom right-hand cell, we see that the average income of the Wave 1 population was £212 per week and that this had risen to £214 per week (both in January 1991 prices) by Wave 2. However, this small increase in average living standards was by no means uniform for different groups in the population. Looking at those who were in the poorest tenth in Wave 1, their average income rose from £84 per week to £107 per week, an increase of a quarter. Even more interesting is the way in which this increase is divided

Table 9.3. Median income in Wave 1 and Wave 2 of individuals classified by decile group in each wave

(£ p.w., January 1991 prices)	In bottom decile group in Wave 2	Not in bottom decile group in Wave 2	All
In bottom decile group in Wave 1	£83 → £85	£84 → £138	£84 → £107
Not in bottom decile group in Wave 1	£159 → £77	£232 → £236	£228 → £227
All	£111 → £81	£224 → £229	£212 → £214

between those who remained in the poorest tenth in both waves and those who 'escaped' by Wave 2.

As the results in the top-left cell indicate, the real incomes of those in the bottom tenth in both waves were relatively stagnant, moving from just £83 per week to £85 per week. However, the group who escaped the poorest tenth saw their real incomes rise very substantially from £84 to £138 per week. This latter group included some who moved from unemployment to employment, or who had a change in family circumstances that brought a significant rise in their household income. The fact that the group who 'escaped' the bottom decile numbered around three-fifths of the original group is indicative of the extent to which personal circumstances may fluctuate from one year to the next.

The group who did worst were those who entered the bottom decile having previously been outside it. Their average incomes fell dramatically from £159 to £77—a drop of over 50 per cent. Indeed, the average income of this group ended up so low that the average income of the poorest tenth as a whole in Wave 2 (£81, from row 3, column 1) was actually below that of the poorest tenth as a whole in Wave 1 (£84, from row 1, column 3).

These movements raise a number of issues, including the extent to which the income mobility seen here is merely a temporary phenomenon. In other words, to what extent do those who escape the bottom decile group between Waves 1 and 2 of the survey enjoy permanently improved living standards and to what extent do they subsequently return to their previously low income levels?

In order to examine this sort of question, we provide some summary analyses based on movements over *three* waves of the BHPS—in other words, covering the period 1991–93 inclusive. Because these analyses are highly preliminary, we begin by broadening the focus to look at movements between *quintile* groups across the whole population over time, rather than focusing exclusively on the bottom decile group.

Our sample of BHPS individuals for whom we have valid household income data for all three waves is 10 835 (including children). Figure 9.2 focuses exclusively on those who were in the bottom quintile group in Wave 1 and tracks their destination in Waves 2 and 3.

Between Wave 1 and Wave 2, 849 (or 39 per cent) of the 2167 individuals who began in the bottom quintile group had escaped into a

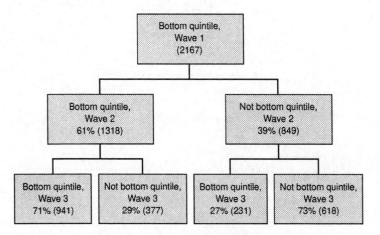

Fig. 9.2. Destinations of bottom quintile group in Wave 1

higher quintile group. In fact, more detailed analysis indicates that more than half of these had escaped only into the second quintile group, but for now we simply subdivide the population into those in the bottom quintile group and those outside it. The remaining 1318 individuals, who form around 61 per cent of the initial bottom quintile group, are also to be found in the bottom quintile in Wave 2.

Continuing to read down Fig. 9.2, we can see that of the 1318 individuals who began in the bottom quintile and remained there in Wave 2, almost three-quarters continued in the bottom quintile in Wave 3. By contrast, of the individuals who had escaped the bottom quintile over the first two waves, only one-quarter subsequently returned to the bottom group, whilst nearly three-quarters sustained their improved position outside the bottom quintile.

Figure 9.3 shows a similar 'transition map' for those who were found outside the bottom quintile group in Wave 1.

In this case, only 10 per cent of those who were outside the bottom quintile fell into the bottom group in Wave 2, and roughly half of these had escaped again by Wave 3. Of the 7819 individuals who remained outside the bottom quintile on both Waves 1 and 2, only 563, or about one in fourteen, fell into the bottom quintile in Wave 3.

A comparison between Figs. 9.2 and 9.3 shows very clearly the way in which an individual's position in the income distribution in a given

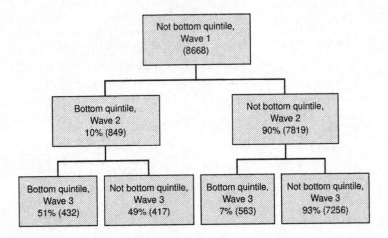

Fig. 9.3. Destinations of those not in bottom quintile group in Wave 1

period is strongly correlated with their position in the previous periods. For example:

(1) an individual who was in the bottom quintile in Wave 1 is *six times* as likely to be in the bottom quintile in Wave 2 as someone who was previously outside the bottom quintile;

(2) an individual who has remained in the bottom quintile for the first two waves is *ten times* as likely to be in the bottom quintile in Wave 3 as someone who has had no previous experience of being in the bottom quintile;

(3) an individual who was in the bottom quintile in Wave 2 but not in Wave 1 is almost *twice* as likely to be in the bottom quintile in Wave 3 as someone who was in the bottom quintile in Wave 1 but not in Wave 2. In other words, the effect of a period in the bottom quintile appears to be more pronounced if it is more recent.

One question that is immediately prompted by these results is 'what distinguishes those whose household incomes are consistently low (or high) and those whose incomes are more volatile?'. Some initial evidence is provided in Table 9.4, which shows the Wave 1 family type of individuals who were in the bottom quintile in Wave 1, subdivided according to whether they remained in the bottom quintile throughout or 'escaped' the bottom quintile at some point.

Table 9.4. Characteristics of individuals remaining in bottom quintile in all waves and of individuals escaping at some point

Wave 1 family type	Of those permanently in bottom quintile (%)	Of those who escaped at some point (%)
Couple pensioner	11	10
Single pensioner	14	14
Couple with children	40	38
Couple, no children	4	13
Single with children	24	12
Single, no children	6	13
Total	100	100

A number of interesting patterns emerge from Table 9.4. First, the group with the highest risk of remaining in the bottom quintile group for three consecutive waves was those who began as lone parents, since this group made up 24 per cent of those who remained on low income but only 12 per cent of those who escaped. This is perhaps to be expected, since a lone parent with a young child has little scope for a significant increase in income except through starting to receive child maintenance or through forming a new partnership.

The group with the highest relative propensity to escape is childless non-pensioner couples, who made up just 4 per cent of those trapped on low incomes compared with 13 per cent of the escapers. Similarly, single childless non-pensioners who began the period on a low income were much less likely to remain poor than families with children or those over pension age. Given that the labour market offers one of the major routes out of low income, it seems reasonable that those with fewest obstacles to employment, such as those without dependent children, are the most likely to see their incomes rise.

Clearly, this initial analysis from just three waves of the BHPS should be regarded as highly preliminary. For example, further waves will provide more information on the extent to which those who escape the bottom quintile group in one year do so only to return a year or two later. Only once we have a significant run of data will it be possible to draw stronger conclusions about the extent of income mobility in Great Britain. None the less, it is clear that 'the poor' are not simply a static group with stable or gradually declining living standards. This analysis also suggests that, when designing anti-poverty policy, it is important to consider the dynamics of poverty—why people become poor, how long they remain poor, and how they cease to be poor—rather than concentrating exclusively on levels of social security benefits.

Some Examples of Other UK Panel Data

The Family Finances Survey

An example of a two-wave panel study undertaken in Great Britain at the end of the 1970s was the Family Finances Survey (FFS).[2] This was a survey of around 3200 low-income families with dependent children carried out between October 1978 and September 1979, with the same families being interviewed again a year later.[3] A total of around 2700 families responded to both waves of the survey.

The original sampling frame for the FFS was families with 'net resources' below 140 per cent of the Supplementary Benefit (SB) level. This was the cut-off point for inclusion in the 'Low Income Families' statistics that were at that time produced by the DHSS. Net resources were family incomes after tax and after the deduction of housing costs.

To some extent, the results from the FFS and its follow-up mirror those obtained from the BHPS as discussed above. Thus, of those families included in the original sample, around one-third had seen their net resources rise to more than 140 per cent of SB levels by a year later, with a fifth attaining net resources above 160 per cent. This suggests a considerable degree of income mobility among the poorest families.

A second interesting finding from the FFS was that a quarter of those who began the period with the lowest recorded living standards (that is, net resources of less than 50 per cent of SB levels) ended the period with net resources in excess of 200 per cent of SB levels—a more than fourfold increase in real income. This suggests that, in some cases, a very low income may be only a temporary state and may not be a good guide to the medium-term prospects of a particular family. This insight is of considerable topical relevance when the very

---

[2] See Hancock (1985) for an extensive analysis of this survey.

[3] The follow-up survey was called the Family Resources Survey. In order to avoid confusion with the Department of Social Security's new and quite separate Family Resources Survey, we do not use the name in the text.

latest low-income statistics are heavily influenced by the presence of growing numbers of households with very low or negative incomes.

It is important, however, not to exaggerate the extent of income mobility or to forget that there is at any time a significant group whose living standards may change relatively little from one year to the next. Analysis of the FFS indicates, for example, that in Wave 2, around one-third of these low-income families had relative resources (expressed as a percentage of the relevant SB line) within 10 percentage points of their Wave 1 level.

DSS Cohort Study of Newly Unemployed Men and Women

In 1987, the Department of Social Security interviewed around 3000 men and women who had become unemployed in the spring of that year. Respondents were asked to provide details of their previous employment and earnings, together with their current income. Nine months later, the same individuals were interviewed again to see what had happened to their employment status and incomes over the intervening period. Around 2100 members of the original sample responded to the follow-up survey and the results provide interesting insights into income dynamics amongst this particular group.

A first interesting finding of the study (reported in Garman, Redmond, and Lonsdale (1992)) was that almost two-thirds of those originally interviewed had had a previous experience of signing on over the preceding five years. In terms of income levels, 53 per cent of the sample had experienced a fall in household income of more than half on becoming unemployed. These two findings indicate that this group had recently experienced considerable volatility in its personal circumstances.

The results of the second interviews indicate that three in five had returned to work at some point in the months following the initial interview, but of these, almost half had since left work again because the employment was of a temporary or seasonal nature. Amongst those who remained unemployed for between 6 and 11 months, the study found that roughly half the sample was more than 10 per cent *better off* than when they had first become unemployed. This somewhat surprising result is largely explained by the benefit disqualifications and limits to mortgage assistance that apply in the early weeks of unemployment but that do not affect those who have been out of work for more than 6 months. Even amongst the continuously unemployed, therefore, income is not as stable over time as might be expected.

Amongst those who moved from unemployment to employment, a not surprising 97 per cent were better off. There were, however, marked differences between men and women, with 63 per cent of men more than doubling their household income on return to work compared with 36 per cent of women. The very large shifts in income associated with transitions into and out of employment again point to an income volatility for certain groups that is not captured by simple cross-section figures for average incomes in a particular group.

DSS Administrative Data

A further source of information about trends in the income levels of particular individuals over time is that routinely collected by the Department of Social Security for the purposes of administering the benefits system. An example of an application of such data is work by Walker (described in Walker with Ashworth (1994)) which used monthly administrative data on flows onto and off Family Credit to examine whether the benefit was becoming a long-term subsidy to low pay. By tracking individuals over time, the study was able to draw the preliminary conclusion that '. . . for most claimants, Family Credit functions as a transitional benefit, bridging families across a short-lived set of circumstances, not as a form of long-term wage subsidy' (p. 199).

'PSEUDO-COHORTS' FROM TIME-SERIES DATA

As we have seen, where panel data have been available for many years, it is possible to examine how the incomes of different sorts of households vary over time. In the UK, the BHPS has been running since 1991 and hence, at the time of writing, only three waves are in the public domain. This raises the question as to whether we can say anything at all about the longer-term movements of individual incomes in the UK.

Before examining in detail one method for getting round the problem of the lack of panel data, it is worth considering why we might be interested in tracking incomes over a longer period, such as an individual's lifetime. One key reason is that the policy implications of low income are likely to be very different according to the stage of life that a given individual has reached.

For example, consider the case of two low-income households in our cross-section survey, one containing a recent school-leaver in a low-paid job and the other containing an elderly pensioner. Historically, the school-leaver might be expected to stay in employment, acquire skills and experience, and gradually see their real income rise. In this case, provided the initial level of income was adequate, there might be relatively little case for state income transfers. To the extent that state intervention was appropriate, it might be in the form of policies designed to ensure that the individual concerned did indeed receive training that would help to improve future income prospects, or intervention in the capital market to make it easier for the individual to smooth living standards over the life cycle.

In the case of the elderly pensioner, the policy response might be quite different. Once someone is well past retirement, there are relatively few sources from which they could expect to receive a significant increase in their income. Instead, their living standard will probably be relatively flat, if not declining as inflation erodes the real value of inadequately indexed private pensions. If the income of an elderly pensioner appears to be low, then there is much more of a case for state intervention, since at this stage in the life cycle there is relatively little prospect of it rising. It is partly for this reason that the means-tested benefit system pays significantly larger sums to individuals aged 80 or over.

Given the importance of considering the life-cycle perspective on individuals' living standards, is there any way in which we can examine the life-cycle pattern of incomes without waiting for sixty years' worth of panel data to become available? In this section, we consider one approach—the use of successive cross-sections of survey data to create 'pseudo-cohorts' of the population.

In order to understand the notion of a 'pseudo-cohort', consider the survey data at our disposal. Family Expenditure Survey data are available at the individual level for every year since 1961. This means we have 33 successive cross-sections with information on a total of around 200000 households. Now suppose we want to look at income dynamics during retirement. In the 1961 survey, the men born in 1896 will have just reached the state pension age of 65. In the 1962 survey, there will be a set of people also born in 1896 who will be aged 66. They will be a different set of people from the 65-year-olds interviewed in 1961, but they may well have similar characteristics. By repeating this process for each successive year of data, we can track the changing circumstances over time of those born in 1896. Note that this group is referred to as a 'pseudo-cohort' because each year's data include different people born in 1896 rather than the same people tracked over time.

Figure 9.4 illustrates the technique applied to every head of benefit unit born in the years 1892–96 over our thirty-three years of data. We have pooled together everyone born over a five-year period in order to achieve a better sample size.

The first thing to notice about Fig. 9.4 is that the mean income of this 'pseudo-cohort' is somewhat erratic. This reflects the fact that the sample includes a different group of people each year, and in particular the fact that the early Family Expenditure Surveys have much smaller samples than those from 1968 onwards.

The general trend shown by the chart is of a rise in average real incomes for this cohort. In part, this simply reflects periodic increases in the real level of the basic state pension. More interestingly, the upward trend is partly explained by the fact that the size of the group is not constant. As the years go by, members of this cohort gradually

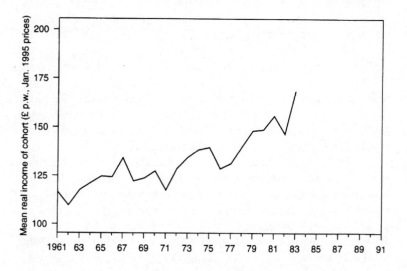

Fig. 9.4.   Mean household equivalent income of 1892–96 cohort

die off, and so by the 1981 survey, for example, the sample consists of those who have reached their eighty-fifth birthday. If, as is the case, those with higher incomes at retirement also have a greater life expectancy subsequent to retirement, then as poorer pensioners die, the average income of the remaining members of the cohort will increase. This is not to say that any of them are actually getting better off, simply that the average income of the cohort as a whole has risen.

The problem of declining sample size due to death does not, of course, significantly affect our examination of younger cohorts. Figure 9.5 shows the average income of those born in the period 1937–41, a cohort where everyone is aged 20 or over at the start of our data.

Figure 9.5 indicates that those who were in their early 20s at the start of the 1960s saw gradual real income growth over the following fifteen years or so, but that by the time they had reached their 40s, they began to experience extremely rapid real income growth. One possible explanation is that people in their 40s may be achieving senior posts which boosts their pay, or perhaps that at this age increasing numbers of the wives of the married men in this group were returning to the labour market and thereby bolstering household income. Alternatively, it could simply have been the case that they

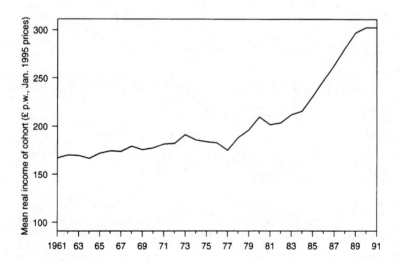

Fig. 9.5. Mean household equivalent income of 1937–41 cohort

were in their 40s at a time when the economy was booming (that is, in the mid- to late 1980s).

One way to distinguish these two effects is to repeat Fig. 9.5 but this time to express the mean income of this group as a fraction of the national average at that time. This is done in Fig. 9.6.

Figure 9.6 indicates that the relative position of this group was somewhat different from that implied by their absolute mean incomes. Although this cohort had average income above the national average throughout the period, there is a distinct U shape to their relative position. During the 1960s, this group saw its position fall back relative to the rest of the population. This is at least in part likely to be attributable to the arrival of children, which increases the number of mouths to feed whilst typically reducing the number of workers in the household. The incomes of this group do, however, gain ground in the late 1970s and forge ahead in the 1980s. This shows that even in the context of the boom years of the late 1980s, this cohort was enjoying a particularly rapid rise in living standards.

Another way of looking at income dynamics is to investigate how the experiences of successive cohorts have changed over time. For

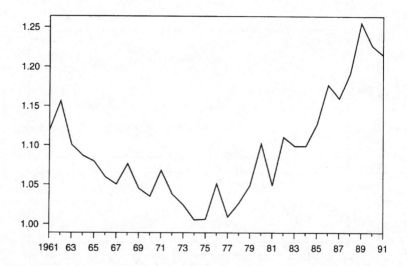

Fig. 9.6. Mean household equivalent income of 1937–41 cohort, expressed as a fraction of the contemporary mean

example, we can pose the question 'was it just the 1937–41 cohort that experienced an income boost in their 40s, or is this just part of the life-cycle pattern and hence something that all groups can expect to experience on average during their 40s?'.

In order to investigate this question, we repeat the analysis of Fig. 9.6 for all cohorts that are at some point aged between 30 and 50 in our data. In this case, however, we plot the age of the cohort (rather than the year of data) along the x-axis in Fig. 9.7. This makes it possible to compare the experience at particular ages of successive cohorts.

A number of interesting patterns are shown in Fig. 9.7. The first is that for most cohorts where data are available over the majority of the period, the path of relative incomes is in the shape of a 'tick', with a small decline in relative household incomes (possibly coinciding with the birth of a child) followed by a substantial and prolonged rise in relative income.

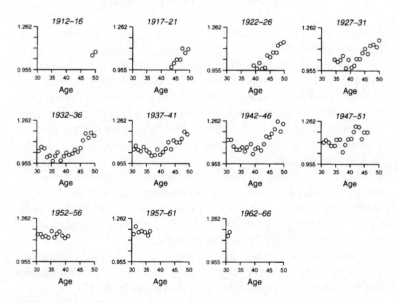

Fig. 9.7. Mean household equivalent income of all cohorts whilst aged 30–50, expressed as a fraction of the contemporary mean

There are, however, exceptions to this general pattern, and in particular the later cohorts seem to be sustaining a relatively high income throughout their early and mid-30s. This may in part reflect a trend towards having children at a later age, but is worthy of more detailed investigation.

There is, therefore, no particular reason to expect the exact profile of life-cycle incomes to remain the same for successive generations. A wide range of factors such as trends in family formation and dissolution, changes in average education levels, and changes in the demand for skills will affect the expected life-cycle income profile of each cohort. None the less, it remains important to bear in mind when examining cross-section descriptions of the income distribution that each cross-section is simply a collection of people at different stages in their life cycles, and that the needs of different individuals with the same income will depend on the stage of the life cycle that they have reached.[4]

SUMMARY AND CONCLUSIONS

Much of the analysis of this book has been based on snapshots of the UK income distribution at particular points in time. These snapshots are, however, the product of dynamic processes that need to be understood in order to interpret trends in the income distribution. In this chapter, we have considered some of the sources of evidence about the dynamics of the income distribution from one year to the next and also about the longer-term patterns of incomes over the life cycle.

It is clear even from the limited evidence available for the UK that an individual's chance of experiencing a relatively low income in a given period is strongly related to income in the previous periods. In particular, in the BHPS data, an individual who has been in the bottom quintile group in the first two waves is ten times as likely to be in the bottom quintile group in the third wave as someone who had not previously been in the bottom quintile group.

Evidence from US panel data suggests that it is helpful to think of a variety of experiences of poverty, from transient, through recurrent and persistent, to permanent. Panel data make it possible to assess

---

[4] See Banks, Blundell, and Preston (1991) for a discussion of the implications of life-cycle considerations in the related context of the construction of equivalence scales.

which forms of poverty are affecting which groups in society and to design public policy responses accordingly.

The lack of long-standing panel studies in the UK makes similar analysis difficult at present, but some idea of the pattern of incomes of particular cohorts can be obtained by the creation of 'pseudo-cohorts' based on successive cross-sections of household survey data. These highlight the profile of living standards for particular cohorts over their life cycle, and also provide some indication that a change in the pattern of lifetime incomes may be occurring, with a rapid rise in relative incomes occurring at an earlier age.

All of these different perspectives indicate that the income distribution should be viewed as the product of a range of dynamic forces rather than simply as a static account of a relatively stable set of personal circumstances. This will inevitably increase the complexity of any analysis, but as more extensive panel data become available, it should be possible to gain a much deeper understanding of the forces that are driving trends in the income distribution.

Conclusions

The last nine chapters have painted a comprehensive picture of the distribution of income (and spending) in the UK. This final chapter draws together some of the threads from that analysis and sets out what seem to be some of the most important conclusions.

THE DEGREE OF INEQUALITY

Interpreting figures showing the distribution of income is difficult unless there is something with which to compare them. What constitutes 'a lot' of inequality and what constitutes a little obviously depend on the criteria against which they are being judged, be they moral or political criteria or merely measured against other distributions at a different time or place. So to look at the income distribution in the UK in the mid-1990s and say that it is very widely dispersed, while being a statement that it is easy to make, is a normative as much as a purely descriptive statement. We have shown that the richest 10 per cent receive a quarter of total income between them while the poorest 10 per cent receive 'just' 3 per cent of total income. The ninetieth percentile to fiftieth percentile ratio is slightly over 2 and the ninetieth to tenth percentile ratio is 4.2. We have presented numerous other similar statistics.

But how unequal does this mean the distribution of income is? Again, the question must be 'unequal compared with what?'. The easiest comparison is with the situation in the past, and compared with the situation ten, twenty, or thirty years ago, one can say that the distribution is more unequal. For one thing that can be stated without fear of contradiction is that the income distribution has grown wider. This rise in inequality appears to have begun towards the end of the 1970s, and to have been particularly speedy in the second half of the 1980s. There has at least been some slowing in that rate of increase during the 1990s, though no clear sign of a major reversal.

All the evidence suggests that the experience of the 1980s is historically highly unusual, if not unprecedented. In terms of the experience of the period since 1960 for which we have consistent data, the sharp rise in inequality over the 1980s contrasts with the relative stability of the two earlier decades. Available official statistics suggest that that period of relative stability stretches back at least until the end of the last war. Indications from yet earlier decades are scarce and probably unreliable, but there is evidence, for example, that the wage distribution among manual workers was quite stable over a long period and that the income share of the very top percentiles fell quite sharply over the middle years of the twentieth century.

Kuznets (1955) hypothesised that, following an initial widening of the income distribution, the disparities in living standards in developed countries would stabilise or move towards greater equality. The last twenty years certainly appear to have consigned that theory to the dustbin of economic history. Looking at the US, the UK, and Germany, Kuznets interpreted the data available in 1955 as showing a movement towards greater equality 'with these trends particularly noticeable since the 1920s but beginning perhaps in the period before the first world war'. That process has gone into sharp reverse in the UK (and in the US).

OTHER MEASURES

Income alone does not tell the whole distributional story, as was explained in Chapter 1 on methodology. One alternative is consumption, or spending. And in Chapter 4, we considered the distribution of spending. This confirmed the picture of increasing inequality over the 1980s but added an extra dimension to our understanding of that increase. The spending distribution did widen, but it did so rather less dramatically than the distribution of income. Perhaps the most important difference is that spending at the bottom of the distribution unambiguously increased in real terms over the period from 1979. Between the early 1980s and early 1990s, the spending levels of the bottom (expenditure) decile group rose by 13 per cent. Incomes of the lowest income group did not rise at all, and on an after-housing-costs basis they actually fell.

Also revealed was the extent to which a large proportion of those at or near the bottom of the income distribution have spending some way up the expenditure distribution. This was true almost exclusively of low-income non-pensioners, most especially of the self-employed, but also of those not in work.

Far from negating the value of the picture drawn from consideration of the income distribution, these facts suggest that one cause of the measured increase in income dispersion is a greater volatility in incomes. People with low incomes for a relatively short period of time are likely to maintain their spending levels above these low, but transient, income levels. Groups with stable incomes, such as pensioners, will not see such variations.

Further evidence of the importance and extent of this potential volatility was examined in Chapter 9 on dynamics, where we showed, using panel data only recently available, that a significant proportion of those in the poorest income group in one year are not to be found in that poorest group the next year. Of those in the poorest quintile in one year, around a half spent at least one of the following two years outside of the bottom quintile. So while only around 10 per cent of the population were 'stuck' in the bottom quintile over these three years, around 30 per cent had some experience of being there. We do not have comparable data for a period prior to the early 1990s, but this does appear to represent a substantial degree of mobility and volatility. The income distribution is not just unequal relative to the past, but it also seems to be rather mobile.

This mobility is of particular importance when considering poverty. A continued experience on a very low income will be very different from a transitory experience. There is a qualitative difference between a population in which everybody spends 10 per cent of their life 'in poverty' and one in which 10 per cent of the population spend their whole life in poverty. Yet standard cross-sectional analyses cannot generally distinguish between such situations.

While considering the issue of poverty in a separate chapter, we cannot provide a definitive answer to the question 'how many people are in poverty?' because, as we discuss, there is no single generally accepted definition of poverty. But it is clear that on any relative measure—based on numbers below a proportion of the contemporary mean income—there was a large increase over the 1980s. There were also many more people dependent on the minimum means-tested benefit in the early 1990s—a near doubling over the period. Despite this, and despite the real reductions in recorded incomes of those at the very bottom of the distribution, one cannot draw an entirely bleak picture. Ownership of all consumer durables among the poorest rose very significantly between 1979 and the early 1990s. More than half of the poorest decile own a car, three-quarters have central heating, and more than two-thirds own a video cassette recorder.

Again the value of looking at measures other than income is apparent; one cannot really understand the changes in living standards of the poor without reference to such measures. The main point about the poor, though, is that there can be no doubt that there is a bigger gap between people at the bottom of the pile and the rest of the population in the 1990s than was the case during the 1960s and 1970s.

WHY THE CHANGE?

These facts lead one quite naturally to the question that has been at the core of this book—'what lies behind this unprecedented increase in inequality?'. It is a simple question without a simple, or a single, answer.

One obvious answer—that it is down to the rise in unemployment—turns out to be only a small part of the story. The multifaceted nature of the inequality increase is perhaps best illustrated in Figs. A and B. They split the population by family type and by economic status, respectively, and, indexing at 100 in 1977, show how the Gini coefficient rose for each family type and for each economic status group in the period 1977–92.

So while there was some divergence in the average income levels experienced by each group, and the relative sizes of the various groups changed substantially over time, an important part of the explanation for the overall increase in inequality must be sought in

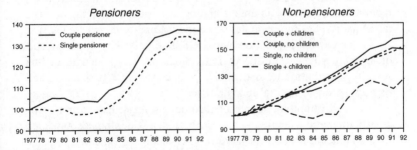

Fig. A. Gini coefficients for each family type

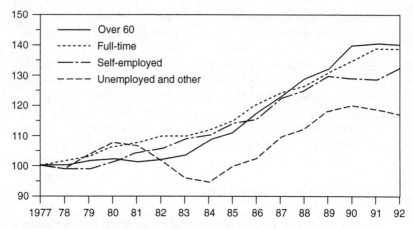

Fig. B.   Gini coefficients for each economic status type

the divergence of experiences within particular groups. In other words, it is not just the increased numbers of unemployed and the increased gap between the incomes of those in and out of work which are responsible for increased inequality, but also the increased gap between well-paid people in work and poorly-paid people in work; between richer pensioners and poorer pensioners; even between the lower-income and higher-income non-pensioners not in work. This, on the basis of a formal decomposition methodology, is the conclusion of Chapter 6.

As for each population sub-group, so for each component of income. Social security apart, each income component now makes a greater contribution to inequality than was the case in the 1970s. That is not quite the same thing as saying that each component is itself more unequally distributed, but it is a more powerful statement when it comes to arriving at an understanding of what is behind the upsurge in inequality. So, for example, while the inequality in private pension receipt across the population actually fell slightly, largely as a result of an increasing proportion of individuals receiving income of this sort, its contribution to inequality rose because both its share in, and its correlation with, total income rose.

The most important source of income, earnings, has seen a particularly dramatic widening in its distribution, a widening accompanied by a significant drop in its share in total income. In the early 1960s, earnings accounted for more than three-quarters of personal income.

This had fallen to 72 per cent by the mid-1970s and then much more dramatically to 61 per cent by the early 1990s. Earnings became concentrated in fewer households, and among those in work the earnings distribution widened dramatically. The largest component of this widening was in the increased inequality of real hourly wages among men. Between 1978 and 1993, the tenth percentile of the male hourly wage distribution barely changed. The median grew by 30 per cent and the ninetieth percentile by more than 50 per cent.

All this and much more is discussed in detail in Chapter 5, which also throws up the one area in which there was actually a reduction in inequality over the 1970s and 1980s—that is, in the relative positions of men and women in the labour market. Their rates of participation in paid work and their hourly wages have both undergone a degree of convergence.

The other set of issues that we consider relate to an area in which government has a degree of direct control over the distribution of income—namely, the tax and social security system. Increase higher rates of income tax and one of the effects will be to reduce the degree of (post-tax) income inequality. Naturally, therefore, the significant changes to the tax system that have been introduced, especially since 1979, have to be considered as a potential culprit behind the widening in the income distribution.

If the tax system had not changed since the late 1970s, then, other things being equal (which, of course, they would not have been), the distribution of net income in the 1990s would be much less unequal than it in fact is. If income tax rates had stayed the same and allowances and bands had just kept pace with inflation, the rich would be paying a great deal more in tax than in fact they do under the current system. But the progressivity of the tax system depends not only on the structure of the system itself, but also on the pre-tax distribution on which it has to work. With even a mildly progressive structure, a widening of the pre-tax distribution will result in an increase in measured progressivity as the top percentiles pay a higher proportion of their income in tax. To a large extent, the widening of the pre-tax income distribution and the reduction in the progressivity of the tax *structure* have offset each other such that the 1990s' tax system has a similar inequality-reducing effect on the 1990s' income distribution to that of the 1970s' tax system on the 1970s' income distribution.

ASSESSING THE EVIDENCE

Clearly, many changes have driven the altered income distribution. A number of things happened in the UK economy at much the same time, particularly during the 1980s. Unemployment rose, the wage dispersion increased, the tax system was reformed. Inequality within virtually every group, however defined, rose. The widening of the income distribution was multifaceted.

But by getting this far, how much have we explained? We have got one step in by identifying, for example, increased wage inequality as part of the explanation. That begs the question 'why did wage inequality increase?'. We can get another step in by identifying a differential experience between the relatively skilled and educated who did well and the unskilled and poorly educated who did badly. Again that begs the question 'why?'. To which the answer lies partly in the changing patterns of demand for different skills, in the role of international competition and of new technology, and in the weakening of institutions such as trade unions and Wages Councils. Of these, not even international competition can be dismissed as a given because its effects are driven by the way in which trade policy is conducted. Certainly, one is still left looking for other economic, political, and social reasons for the changing returns to skills and for the policies that led to the relative demise of the labour market institutions.

Much the same arguments apply to unemployment. The rise in the numbers of people not in work was one of the events shaping the income distribution over the 1980s. Also important was the distribution of workers among households—a decline in the number of single-earner couples and an increase in the numbers of no-earner and two-earner couples. These changes were shaped by many of the same things that affected the wage distribution, by government policies, by wider economic change, and by social change.

A similarly long path of enquiry beckons when considering the experience of pensioners, though there may be fewer imponderables at the end of the path. The rise in inequality among pensioners reflects, as we saw, the divergent experience of those with and without private sources of income. There has been a rise in levels of occupational pensions but many pensioners are still dependent on state benefits. This rise in private income levels is readily traceable to increased pension coverage in the work-force during the 1950s and

1960s and improved legislative protection in the 1980s. Increasingly, divergent private pension incomes will arise from divergent labour market experiences, the multiplicity of reasons for which we have just considered. But the experience of inequality increase would have been nothing like so high had it not been for the fact that state benefits for pensioners were largely pegged to prices. This policy was itself a result of a range of political factors and decisions. As with changes to other parts of the tax and social security systems, the impact of the change is relatively easily measurable but the reasons for it are rather more complex, lying in the field of politics as much as economics.

CONCLUSIONS

A natural question to ask at this point is 'what will happen next?'. Have we reached a plateau in inequality, will it continue rising, or are we going to return to a distribution more like that experienced over most of the post-war period? And if inequality is not going to fall of its own accord, and if we wanted it to fall, is there anything to be done to nudge it in that direction?

Crystal-gazing is a dangerous habit, especially when the results are printed and published. But it does seem to us unlikely that over the next few years, we will see a return to an income distribution closer to that of the 1960s or 1970s than to that of today. An economic and political upheaval at least as great as that experienced during the 1980s would be required for such a reversal to occur. And it would be an upheaval that would be counter to many of the global economic imperatives that helped drive the experiences of the 1980s.

It is rather more difficult to predict whether the current distribution will remain much as it is or whether we will continue to see a further widening in the dispersion. If we did want to make predictions, we would have to take a view on likely trends in the labour market, on both supply and demand sides, on the political climate and political decisions, and on the future direction and impact of technological and global economic change.

One of the most important changes of the 1980s was the relative decline in the demand for less-skilled workers resulting in both relatively lower wage levels and a lower likelihood of being in employment. If this continues, it will be a powerful force for a continued increase in inequality. Continued mechanisation and globalisation point in this direction. Possible countering forces include the fact that we must be approaching a point at which the supply of female labour will stop increasing as quickly as it has done, as should the supply of unskilled or inappropriately skilled individuals joining the labour force or being made redundant from 'old' manufacturing industries. A possible continued increase in demand for relatively unskilled labour in the service and 'caring' sectors might also serve to ameliorate some of the forces towards increased unemployment and wage dispersion. It is at least at these sorts of economic forces that we would have to look, very carefully, in order to make some sort of estimate of possible future changes.

On the political side, the choices are over continued deregulation of the labour market and over tax policy and social security policy. A reduction in the power of labour market institutions—trade unions and Wages Councils especially—appears to have contributed to the growth in wage inequality. Political decisions over whether or not to stop or reverse such trends will be of some importance. The introduction of a minimum wage, for example, would significantly alter the institutional framework.

Future decisions over taxation and benefit policies will, of course, be important. A decision to increase benefits in line with earnings would prevent a growth in the gap between workers and non-workers, and between poorer pensioners and their more fortunate counterparts. A continued reliance on price indexation will see these poorer groups falling further behind the bulk of the working population. In the long term, continued earnings growth, especially for those at the top end of the distribution, and continued price indexation of benefits will inevitably result in further massive changes in the distributional picture. The fact is that higher social security benefits and a significantly more progressive tax system appear to be off the political agenda. Nobody has a coherent plan for significantly reducing the level of non-employment. Over the very long run, there is some consensus that education and training are vital, but no clear plans for expanding and improving them.

That there is so little clear urge on the part of politicians of any party to reverse the widening of the income dispersion witnessed over the past decades itself bears witness to an important fact. The way in which income is distributed can, and does, play a major role in determining the outcome of the political process. We return to the arguments set out in the Introduction. Where enough people are, through their higher incomes, divorced from the experiences of the poor, then their political concerns may be less likely to focus on redistribution. And where they even have sufficient income to opt out of parts of the welfare state, their support for that part of the welfare state is likely to be diminished.[1]

And then the wheel has turned full circle. Political choices unquestionably contributed to the rise in inequality. That widening of the distribution itself makes similar political choices easier to sustain. It becomes harder for choices that reverse the changes to gain adequately widespread acceptance. In the absence of powerful countervailing economic forces—and the most powerful economic forces are far from countervailing—a new equilibrium is reached. Like a supertanker that requires vast expanses of ocean in which to slow down and turn, once an increase in inequality has gained momentum, slowing it down and reversing it is likely to take many, many years.

---

[1] Besley, Hall, and Preston (1996) provide evidence for this with respect to private healthcare.

REFERENCES

Abel-Smith, B., and Townsend, P., (1965), *The Poor and the Poorest*, London: Bell.

Ando, A., and Modigliani, F., (1963), 'The life-cycle hypothesis of saving: aggregate implications and tests', *American Economic Review*, vol. 53, pp. 55–84.

Aronson, A., Johnson, P., and Lambert, P., (1994), 'Redistributive effect and unequal income tax treatment', *Economic Journal*, vol. 104, pp. 262–70.

Ashenfelter, O., and Layard, R., (1983), 'Incomes policy and wage differentials', *Economica*, vol. 50, pp. 127–43.

Atkinson, A., (1970), 'On the measurement of inequality', *Journal of Economic Theory*, vol. 2, pp. 244–63.

—— (1983), *The Economics of Inequality*, second edition, Oxford: Clarendon Press.

——, Gomulka, J., and Sutherland, H., (1988), 'Grossing up FES data for tax benefit models', in A. Atkinson and H. Sutherland (eds), *Tax Benefit Models*, London: Suntory Toyota International Centre for Economics and Related Disciplines.

——, Gordon, J., and Harrison, A., (1989), 'Trends in the share of top wealth holders in Britain, 1923–1981', *Oxford Bulletin of Economics and Statistics*, vol. 51, pp. 315–31.

—— and Micklewright, J., (1983), 'On the reliability of income data in the Family Expenditure Survey 1970–1977', *Journal of the Royal Statistical Society*, vol. 146, part 1, pp. 33–61.

—— and —— (1989), 'Turning the screw: benefits for the unemployed 1979–88', in A. Dilnot and I. Walker (eds), *The Economics of Social Security*, Oxford: Oxford University Press.

——, Rainwater, L., and Smeeding, T., (1995), *Income Distribution in OECD Countries*, Social Policy Studies no. 18, Paris: Organisation for Economic Co-operation and Development.

Banks, J., Blundell, R., and Lewbel, A., (1992), 'Quadratic logarithmic Engel curves and consumer demand', Institute for Fiscal Studies, Working Paper no. 92/14.

——, ——, and Preston, I., (1991), 'Adult equivalence scales: a life-cycle perspective', *Fiscal Studies*, vol. 12, no. 3, pp. 16–29.

——, ——, and —— (1994), 'Measuring the life cycle consumption costs of children', in R. Blundell, I. Preston, and I. Walker (eds), *The Measurement of Household Welfare*, Cambridge: Cambridge University Press.

——, ——, and Tanner, S., (1995), 'Is there a retirement-savings puzzle?', Institute for Fiscal Studies, Working Paper no. 95/4.

—— and Johnson, P., (1993), *Children and Household Living Standards*, London: Institute for Fiscal Studies.

—— and —— (1994), 'Equivalence scale relativities revisited', *Economic Journal*, vol. 104, pp. 883–90.

Becker, G., (1981), *A Treatise on the Family*, Cambridge MA: Harvard University Press.

Besley, T., Hall, J., and Preston, I., (1996), *Private Health Insurance and the State of the NHS*, Commentary no. 52, London: Institute for Fiscal Studies.

Blundell, R., and Lewbel, A., (1991), 'The information content of equivalence scales', *Journal of Econometrics*, vol. 50, pp. 49–68.

—— and Preston, I., (1995), 'Income, expenditure, and the living standards of UK households', *Fiscal Studies*, vol. 16, no. 3, pp. 40–54.

Boden, R., and Corden, A., (1994), *Measuring Low Incomes: Self-Employment and Family Credit*, Social Research Policy Unit, London: HMSO.

Borooah, V., McGregor, P., McKee, P., and Mulholland, G., (1996), 'Cost of living differences between the regions of the United Kingdom', in J. Hills (ed.), *New Inequalities*, Cambridge: Cambridge University Press.

—— and McKee, P., (1995), 'How much did working wives contribute to changes in income inequality between couples in the UK?', *Fiscal Studies*, vol. 17, no. 1, pp. 59–78.

Bradshaw, J., (1993), *Budget Standards for the United Kingdom*, Aldershot: Avebury.

Buck, N., Gershuny, J., Rose, D., and Scott, J., (1994), *Changing Households: The British Household Panel Survey, 1990–1992*, Colchester: ESRC Research Centre on Micro-Social Change, University of Essex.

Burtless, G., (1995), 'International trade and the rise in earnings inequality', *Journal of Economic Literature*, vol. 33, pp. 800–16.

Card, D., (1991), 'The effect of unions on the distribution of wages: redistribution or relabelling?', Princeton University Industrial Relations Section, Discussion Paper no. 287.

Clarke, A., and Oswald, A., (1994), 'Unhappiness and unemployment', *Economic Journal*, vol. 104, pp. 648–59.

Clegg, H., (1979), *The Changing System of Industrial Relations in Britain*, Oxford: Basil Blackwell.

Corry, D., and Glyn, A., (1994), 'The macroeconomics of equality, stability and growth', in A. Glyn and D. Miliband (eds), *Paying for Inequality: The Economic Cost of Social Injustice*, London: Rivers Oram Press.

Coulter, F., Cowell, F., and Jenkins, S., (1992), 'Equivalence scale relativities and the extent of inequality and poverty', *Economic Journal*, vol. 102, pp. 1067–82.

Cowell, F., (1995), *Measuring Inequality*, LSE Handbooks on Economics, Hemel Hempstead: Prentice Hall / Harvester Wheatsheaf.

Crawford, I., (1994), *UK Household Cost of Living Indices: 1979–1992*, Commentary no. 44, London: Institute for Fiscal Studies.

CSO (1995), 'The effects of taxes and benefits on household income, 1994/ 95', *Economic Trends*, no. 506, pp. 25–59.

Cutler, D., and Katz, L., (1992), 'Rising inequality? Changes in the distribution of income and consumption in the 1980s', *American Economic Review*, vol. 82, pp. 546–51.

Davies, M., (1995), *Household Incomes and Living Standards: The Interpretation of Data on Very Low Incomes*, Analytical Notes no. 4, London: Analytical Services Division, Department of Social Security.

Deaton, A., and Muellbauer, J., (1980), *Economics and Consumer Behaviour*, Cambridge: Cambridge University Press.

Dilnot, A., and Johnson, P., (1992), 'What pension should the state provide?', *Fiscal Studies*, vol. 13, no. 4, pp. 1–20.

——, Kay, J., and Keen, M., (1989), 'Allocating taxes to households: a methodology', Institute for Fiscal Studies, Working Paper no. 89/11.

——, ——, and Morris, N., (1984), *The Reform of Social Security*, Oxford: Clarendon Press.

Disney, R., Meghir, C., and Whitehouse, E., (1994), 'Retirement behaviour in Britain', *Fiscal Studies*, vol. 15, no. 1, pp. 24–43.

—— and Webb, S., (1991), 'Why are there so many long term sick in Britain?', *Economic Journal*, vol. 101, pp. 252–62.

DSS (1995), *Households Below Average Income 1979–1992/93: A Statistical Analysis*, London: HMSO.

EC (1991), *Final Report of the Second European Poverty Programme*, Brussels: Commission of the European Communities.

Engel, E., (1895), 'Die Lebenskosten Belgischer Arbeiter-Familien frueher und jetzt', *International Statistical Institute Bulletin*, vol. 9, pp. 1–74.

Evandrou, M., Falkingham, J., Hills, J., and Le Grand, J., (1993), 'Welfare benefits in kind and income distribution', *Fiscal Studies*, vol. 14, no. 1, pp. 57–76.

Evans, M., (1995), 'Out for the count: the incomes of the non-household population and the effect of their exclusion from national income profiles', London School of Economics, Welfare State Programme Discussion Paper no. WSP/111.

Feinstein, C., (1996), 'The equalising of wealth in Britain since the Second World War', *Oxford Review of Economic Policy*, vol. 12, no. 1, pp. 96–105.

Ford, J., (1991), *Consuming Credit: Debt and Poverty in the UK*, London: Child Poverty Action Group.

Frayman, H., Mack, J., Lansley, S., Gordon, D., and Hills, J., (1991),

Breadline Britain in the 1990s: The Findings of the Television Series, London: London Weekend Television.

Freeman, R., (1993), 'How much has deunionisation contributed to the rise in male earnings inequality?', in S. Danziger and P. Gottschalk (eds), *Uneven Tides: Rising Inequality in America*, New York: Russell Sage Foundation.

Friedman, M., (1957), *A Theory of the Consumption Function*, Princeton: Princeton University Press.

Fry, V., and Stark, G., (1990), 'New rich or old poor: poverty, take-up and the indexation of the state pension', *Fiscal Studies*, vol. 12, no. 1, pp. 67–77.

Gardiner, K., (1994), 'A survey of income inequality over the last twenty years: how does the UK compare?', London School of Economics, Welfare State Programme Discussion Paper no. WSP/100.

Garman, A., Redmond, G., and Lonsdale, S., (1992), *Incomes In and Out of Work: A Cohort Study of Newly Unemployed Men and Women*, Research Report no. 7, London: Department of Social Security.

Giles, C., and Johnson, P., (1994), 'Tax reform in the UK and changes in the progressivity of the tax system, 1985–95', *Fiscal Studies*, vol. 15, no. 3, pp. 64–86.

Goodman, A., Johnson, P., and Webb, S., (1994), 'The UK income distribution 1961–1991: the role of demographic and economic changes', paper presented at the International Association for Research into Income and Wealth conference, St Andrews, Canada, 21–27 August.

—— and Webb, S., (1994), 'For richer, for poorer: the changing distribution of income in the UK, 1961–91', *Fiscal Studies*, vol. 15, no. 4, pp. 29–62.

—— and —— (1995), 'The distribution of UK household expenditure, 1979–92', *Fiscal Studies*, vol. 16, no. 3, pp. 55–80.

Gosling, A., and Machin, S., (1995), 'Trade unions and the dispersion of earnings in British establishments', *Oxford Bulletin of Economics and Statistics*, no. 57, pp. 167–84.

——, ——, and Meghir, C., (1994), 'What has happened to men's wages since the mid-1960s?', *Fiscal Studies*, vol. 15, no. 4, pp. 63–87.

Green, A., (1994), *The Geography of Poverty and Wealth: Evidence on the Changing Spatial Distribution and Segregation of Poverty and Wealth from the Census of Population, 1991 and 1981*, Warwick: Institute of Employment Research.

Gregg, P., and Wadsworth, J., (1996), 'More work in fewer households', in J. Hills (ed.), *New Inequalities*, Cambridge: Cambridge University Press.

Hancock, R., (1985), 'Explaining changes in families' relative net resources: an analysis of the Family Finances and Family Resources Surveys', London School of Economics, TIDI Working Paper no. 84.

—— and Weir, P., (1994), *More Ways than Means: A Guide to Pensioners' Incomes in Great Britain during the 1980s*, London: Age Concern Institute of Gerontology, King's College.

Harkness, S., Machin, S., and Waldfogel, J., (1996), 'Evaluating the pin money hypothesis: the relationship between women's labour market activity, family income, and poverty in Britain', in J. Hills (ed.), *New Inequalities*, Cambridge: Cambridge University Press.

Harris, G., and Davies, M., (1994), *Income Measures for Official Low Income Statistics: The Treatment of Housing Costs and Local Government Taxes*, Analytical Notes no. 2, London: Analytical Services Division, Department of Social Security.

Hills, J., (1995), *Inquiry into Income and Wealth. Volume 2: A Summary of the Evidence*, York: Joseph Rowntree Foundation.

Hutton, S., (1995), 'Men's and women's incomes: evidence from survey data', *Journal of Social Policy*, vol. 23, pp. 21–40.

Hutton, W., (1995), *The State We're In*.

Jenkins, S., (1991), 'Income inequality and living standards: changes in the 1970s and 1980s', *Fiscal Studies*, vol. 12, no. 1, pp. 1–28.

—— (1994a), 'The within household distribution and why it matters: an economist's perspective', University of Swansea, Economics Discussion Paper no. 94–05.

—— (1994b), 'Winners and losers: a portrait of the income distribution in the UK over the 1980s', University College of Swansea, Economics Discussion Paper no. 94–07.

—— (1995), 'Accounting for inequality trends: decomposition analyses for the UK, 1971–86', *Economica*, vol. 62, pp. 29–63.

—— and Cowell, F., (1994), 'Dwarfs and giants in the 1980s: trends in the UK income distribution', *Fiscal Studies*, vol. 15, no. 1, pp. 99–118.

—— and O'Leary, N., (1994), 'Household income plus household production: the distribution of extended income in the UK', University of Swansea, Economics Discussion Paper no. 94–14.

Johnson, P., and Stark, G., (1989), 'Ten years of Mrs Thatcher: the distributional consequences', *Fiscal Studies*, vol. 10, no. 2, pp. 29–37.

—— and —— (1991), 'The effects of a minimum wage on family incomes', *Fiscal Studies*, vol. 12, no. 3, pp. 88–93.

—— and Stears, G., (1995), 'Pensioner income inequality', *Fiscal Studies*, vol. 16, no. 4, pp. 69–94.

—— and Webb, S., (1992), 'The treatment of housing in official low income statistics', *Journal of the Royal Statistical Society*, no. 155, part 2, pp. 273–90.

—— and —— (1993), 'Explaining the growth in UK income inequality: 1979–1988', *Economic Journal*, vol. 103, pp. 429–35.

Jones, J., Stark, G., and Webb, S., (1991), 'Modelling benefit expenditures using the FES', Institute for Fiscal Studies, Working Paper no. 91/13.

Joseph Rowntree Foundation (1995), *Inquiry into Income and Wealth. Volume 1*, York: Joseph Rowntree Foundation.

Juhn, C., Murphy, K., and Pierce, B., (1993), 'Wage inequality and the rising returns to skill', *Journal of Political Economy*, vol. 101, pp. 410–42.

Kemsley, W., Redpath, R., and Holmes, M., (1980), *Family Expenditure Survey Handbook*, London: HMSO.

Kuznets, S., (1955), 'Economic growth and income inequality', *American Economic Review*, vol. 45, pp. 1–28.

—— (1963), 'Quantitative aspects of the economic growth of nations: part VIII, distribution of income by size', *Economic Development and Cultural Change*, vol. 11.

Lambert, P., (1989), *The Distribution and Redistribution of Income: A Mathematical Analysis*, Oxford: Basil Blackwell.

Layard, R., Piachaud, D., and Stewart, M., (1978), *The Causes of Poverty*, Royal Commission on the Distribution of Income and Wealth, Background Paper to Report no. 5, London: HMSO.

Lazear, E., and Michael, R., (1988), *Allocation of Income within the Household*, Chicago and London: University of Chicago Press.

McClements, L., (1977), 'Equivalence scales for children', *Journal of Public Economics*, vol. 8, pp. 191–210.

Machin, S., (1996), 'Wage inequality in the UK', *Oxford Review of Economic Policy*, vol. 12, no. 1, pp. 47–64.

—— and Manning, A., (1994), 'Minimum wages, wage dispersion and employment: evidence from the UK Wages Councils', *Industrial Labour Relations Review*, no. 47, pp. 319–29.

Mack, J., and Lansley, S., (1985), *Poor Britain*, London: Allen and Unwin.

Meager, N., Court, G., and Moralee, J., (1996), 'Self-employment and the distribution of income', in J. Hills (ed.), *New Inequalities*, Cambridge: Cambridge University Press.

Mookherjee, D., and Shorrocks, A., (1982), 'A decomposition analysis of the trend in UK income inequality', *Economic Journal*, vol. 92, pp. 886–902.

Morris, N., and Preston, I., (1986), 'Inequality, poverty, and the redistribution of income', *Bulletin of Economic Research*, vol. 38, pp. 279–85.

Muellbauer, J., (1979), 'McClements on equivalence scales for children', *Journal of Public Economics*, vol. 12, no. 2, pp. 221–31.

—— (1980), 'The estimation of the Prais-Houthakker model of equivalence scales', *Econometrica*, vol. 48, no. 1, pp. 153–76.

Nicholson, J., (1949), 'Variations on working class family expenditure', *Journal of the Royal Statistical Society*, Series A, vol. 112, pp. 359–411.

Noble, M., Smith, G., Avenall, D., Smith, T., and Sharland, E., (1994), *Changing Patterns of Income and Wealth in Oxford and Oldham*, Oxford: Department of Applied Social Studies and Social Research, University of Oxford.

Nolan, B., and Whelan, C., (1996), *Resources, Deprivation and Poverty*, Oxford: Oxford University Press.

Nozick, R., (1974), *Anarchy, State, and Utopia*, Oxford: Basil Blackwell.

O'Higgins, M., and Jenkins, S., (1990), 'Poverty in the EC: estimates for 1975, 1980 and 1985', in R. Teekens and B. van Praag (eds), *Analysing Poverty in the European Community*, Luxemburg: Eurostat.

Pen, J., (1971), *Income Distribution*, London: Allen Lane.

Pollak, R., and Wales, T., (1979), 'Welfare comparisons and equivalence scales', *American Economic Review*, vol. 69, pp. 216–21.

Pryke, R., (1995), *Taking the Measure of Poverty. A Critique of Low Income Statistics: Alternative Estimates and Policy Implications*, Research Monograph no. 51, London: Institute of Economic Affairs.

Ray, R., (1986), 'Demographic variables and equivalence scales in a flexible demand system: the case of AIDS', *Applied Economics*, vol. 18, pp. 265–78.

Rothbarth, E., (1943), 'Note on a method of determining equivalent income for families of different composition', Appendix 4 in C. Madge, *War Time Pattern of Saving and Spending*, Occasional Paper no. 4, Cambridge: Cambridge University Press.

Rowntree, J., (1901), *Poverty: A Study of Town Life*, 1992 edition, London: Macmillan.

Royal Commission on the Distribution of Income and Wealth (1979), *Report no. 7: Fourth Report on the Standing Reference*, London: HMSO.

Schmitt, J., (1995), 'The changing structure of male earnings in Britain, 1974–1988', mimeo, London School of Economics; forthcoming in R. Freeman and L. Katz (eds), *Changes and Differences in Wage Structures*, Chicago IL: University of Chicago Press.

Sen, A., (1984), *Resources, Values and Development*, Oxford: Basil Blackwell.

—— (1985), 'The standard of living: Lecture I—concepts and critiques', in G. Hawthorne (ed.), *The Standard of Living*, Cambridge: Cambridge University Press.

Shorrocks, A., (1982a), 'Inequality decomposition by factor components', *Econometrica*, vol. 50, pp. 193–211.

—— (1982b), 'The impact of income components on the distribution of family incomes', *Quarterly Journal of Economics*, vol. 98, pp. 311–26.

Slesnick, D., (1992), 'Consumption, needs and inequality', *International Economic Review*, vol. 35, pp. 677–703.

Smeaton, D., and Hancock, R., (1995), *Pensioners' Expenditure: An Assessment of Changes in Living Standards, 1979–1991*, London: Age Concern Institute of Gerontology, King's College.

Social Security Committee (1993), *Low Income Statistics: Low Income Families 1979–1989*, House of Commons, Session 1992–93, London: HMSO.

—— (1995), *Low Income Statistics: Low Income Families 1989–92*, House of Commons, Session 1994–95, London: HMSO.

Theil, H., (1967), *Economics and Information Theory*, Amsterdam: North Holland.

—— (1972), *Statistical Decomposition Analysis*, Amsterdam: North Holland.

Titmuss, R., (1962), *Income Distribution and Social Change*, London: Allen and Unwin.

van Praag, B., Hagenaars, A., and Van Weeren, J., (1982), 'Poverty in Europe', *Review of Income and Wealth*, vol. 28, pp. 345–59.

Walker, R., with Ashworth, K., (1994), *Poverty Dynamics: Issues and Examples*, Aldershot: Avebury.

Webb, S., (1993), 'Women's incomes: past, present and prospects', *Fiscal Studies*, vol. 14, no. 4, pp. 14–36.

—— (1994), 'Social insurance and poverty alleviation: an empirical analysis', in S. Baldwin and J. Falkingham (eds), *Social Security and Social Change: New Challenges to the Beveridge Model*, London: Harvester Wheatsheaf.

Index